T0203027

PATENT LAW
FOR
SCIENTISTS
AND
ENGINEERS

PATENT LAW
FOR
SCIENTISTS
AND
ENGINEERS

EDITED BY
AVERY N. GOLDSTEIN

CRC Press
Taylor & Francis Group
Boca Raton London New York

CRC Press is an imprint of the
Taylor & Francis Group, an **informa** business

CRC Press
Taylor & Francis Group
6000 Broken Sound Parkway NW, Suite 300
Boca Raton, FL 33487-2742

First issued in paperback 2019

© 2005 by Taylor & Francis Group, LLC
CRC Press is an imprint of Taylor & Francis Group, an Informa business

No claim to original U.S. Government works

ISBN-13: 978-0-8247-2383-5 (hbk)
ISBN-13: 978-0-367-39321-2 (pbk)

Library of Congress Cataloging-in-Publication Data

Catalog record is available from the Library of Congress

Visit the Taylor & Francis Web site at
http://www.taylorandfrancis.com

and the CRC Press Web site at
http://www.crcpress.com

Preface

Technology is the watchword of our age. Corporations and universities have responded to changes in the economic order by accelerating the pace of technological development and commercialization. A successful company now must have superior technology and to justify research expenditures, the resulting intellectual property must be protectable.

The patent system represents a bargain between the inventor and society: in exchange for teaching the public of an inventor's discovery, society gives the inventor a limited term monopoly to exclude others from practicing the invention. The tinkers and visionaries of the previous times have largely been replaced by a professional inventor class of scientists and engineers who derive livelihoods by the solving of complex technological problems. The solutions are brought to the benefit of the institution, and ultimately society, only through the efforts of other professionals who take an invention through the complex manufacturing, regulatory, and legal facets of modern society.

The change in invention setting from the romantic notions of a sole inventor toiling through the night to professional scientists and engineers employed in an institutional workplace has affected not only the philosophy and nature of science but the responsibilities charged to the inventor. A scientist or engineer practicing their craft now must be aware of how patent rights are woven through the research process. A breakthrough discovery without proper patent protection may never reach the public, since investment in the discovery cannot be justified if there is not a time of exclusivity to recoup the cost of investment capital. This work has been assembled under the premise that patent rights are integral to the work of the scientist and engineer and not an adjunct to the solution of technical problems.

While many texts have been written to deliver an understanding of intellectual property law to scientists and engineers, these works have generally failed to provide an appropriate scope, which is neither too expansive nor too detailed. Rather than attempt to give a mile-high view of all types of intellectual property or, at the other extreme, to turn the technical reader into a pseudo-patent attorney, this work is intended to provide the practicing scientist, engineer, or student with the understanding of those aspects of patent law that are needed to best protect their inventions. Thus, for the secondary forms of intellectual property from the standpoint of a scientist or engineer, trademark and copyright law, as well as the mechanics of patent prosecution, the reader is generally referred elsewhere.

An assumption is inherent in this volume that the reader will have the benefit of interacting with an information specialist to search patent databases, and a patent agent or attorney to draft and prosecute patent applications. It is strongly recommended that an inventor seek out such patent professionals to assure that a potential invention be afforded the greatest opportunity to obtain the full protection available under the patent laws of various countries and multinational treaties. It is my intention that the reader finds the following pages filled with information that can be implemented into the daily research routine.

Wherever practical, the issues discussed in a given chapter are followed with fact patterns to emphasize actions necessary to protect the latent patent rights that may exist in the solution of a technical problem. The illustration of actual scenarios encountered by an engineer or scientist are intended to highlight a practical course of action to best protect latent patent rights that may well exist in an invention.

Avery N. Goldstein

Editor

Avery N. Goldstein is a partner at the intellectual law firm of Gifford, Krass, Groh, Sprinkle, Citkowski & Anderson, P.C. He is admitted to practice before the United States Patent and Trademark Office, the courts of the State of Michigan and the federal courts. His practice is focused on biotechnology, chemistry, and nanotechnology patent prosecution. A patent he prosecuted was recently named as one of the 10 most important patents in nanotechnology by *Nanotechnology Law & Business*. He was the editor of the *Handbook of Nanophase Materials* (1997). He previously worked as a Senior Research Chemist in the chemical industry. He has authored over 20 papers in the field and holds several patents in the area of nanotechnology. He holds a Bachelor of Science degree in Chemistry, Bachelor of Science in Biological Science, and a Juris Doctor degree from Wayne State University and a doctorate degree in Chemistry from the University of California at Berkeley. He is married with two children.

Contributors

Tom Brody
Registered Patent Agent
Coudert Brothers L.L.P
San Francisco, California

Angela M. Davison
Intellectual Property Counsel
Ross Controls
Troy, Michigan

Ernest I. Gifford
Partner
Gifford, Krass, Groh, Sprinkle,
    Anderson & Citkowski, P.C.
Troy, Michigan

Avery N. Goldstein
Partner
Gifford, Krass, Groh, Sprinkle,
    Anderson & Citkowski, P.C.
Troy, Michigan

Roberta J. Morris
Patent Attorney
Ann Arbor, Michigan

Peter J. Newman
University of Alabama at
    Birmingham
Birmingham Office of Grants and
    Contracts Administration
    (OGCA)
Birmingham, Alabama

Judith M. Riley
Partner
Gifford, Krass, Groh, Sprinkle,
Anderson & Citkowski, P.C.
Troy, Michigan

Contents

Part IV: Ancillary patent activities

Part I

Introduction

chapter one

Anatomy of a patent

Roberta J. Morris
Patent Attorney

Contents

1.1   Introduction

What is a patent? In this chapter we will look at a real patent, and understand the parts of the patent and why they are, and must be, there. We will
also learn a little about the legal effects of having a patent — what a patent
entitles its owners to do, and what it does not entitle them to do.

1.1.1   What patents are, and are not

Here is a practical definition of "patent":

> A patent is a grant
>   from the government
>   to someone who demonstrates to the satisfaction of the
>   Patent Office that
>   something new and useful has been invented,
>   and, once issued, it permits the owner to exclude other people
>   from practicing the patented invention or putting it in the stream
>   of commerce.

There are a few key words in this sentence, and you probably have a feel for
what they mean but you may not be totally sure. Among the first that you
need to understand (and that patent lawyers might use in talking to new
clients, assuming that the phrase is transparent when it might not be) is
"practicing the invention."

Hah! Fooled you! You thought I would focus first on "new" and "useful." Those words are *very* important, true, and they are the subject of this
and later chapters. They are also **terms of art**, which means that they have
special legal significance. Terms of art can mean what you think they mean,
but you should be careful when you use them so that you distinguish
between their colloquial meanings and their legal ones.

"Practicing the invention" is a phrase you may hear your patent
lawyers say and, while they may not doubt that you understand it, you
may not feel so certain yourself. So let us start with "practicing." I practice
an invention if I **do** it — if it is a process or method; or if I **make** it — if it
is a piece of equipment, chemical, drug, or bit of biological matter; *and* if
I do or make it the way the patent says to do or make it. This brings us to
another key concept: how does the rest of the world know what the patent
"says"?

A patent (see the following pages for a sample) speaks in two ways, it **teaches** and it **claims**. Sometimes what your patent "says" is what it teaches, and sometimes it is what it claims. When patent lawyers speak of someone practicing the patented invention, they mean that what the person does is covered by one or more of the *claims* of the patent.

Figure 1 on pages 6 to 11 is a copy of U.S. Patent No. 6,055,695 entitled "Lint Roller Assembly" as issued by the United States Patent and Trademark Office. Figure 1.1 is the cover page. The claims are at the end of the patent (see Figure 1.3b, column 4, lines 4 to 50). The claims are numbered and each must complete a single sentence that begins "What I (or we) claim is...." We discuss claims more fully below.

The important thing to remember is that the word "claim" is a term of art in patent law; it does not have its ordinary meaning. What you claim is not what you argue, or what you contend, or what you assert you want in the lawsuit (all of which are fine synonyms for "what you claim" in other areas of the law, as well as everyday speech). What you *claim* in your patent is (or are) specific sentences that use language to describe your invention, in three dimensions, as it changes in time, as it is put together, as it does its work, etc.

We will discuss the claims in more detail later, but first let us go back to the definition of a patent and examine some of the other words and phrases.

1. How do you put an invention into the stream of commerce? Under the patent statute, you do that by selling it, or offering it for sale, or importing it. Someone who does any of those things without authorization from the patent owner may not himself be practicing the invention, but the patent statute designates those things as infringement too.

2. How do you "demonstrate [anything] to the satisfaction of the Patent Office?" You apply for a patent. This is different from a copyright, but like a trademark (see Table 1.1 on page 14). I use the phrase "Patent Office" but the full name is United States Patent and Trademark Office; it is a part of the United States Department of Commerce. You satisfy the Patent Office by complying with requirements from three main sources:

   a. statutes: enacted by Congress and signed into law by the President;

   b. regulations: promulgated by the Patent Office as a government agency; and

   c. internal guidelines of the Patent Office, such as the Manual of Patent Examining Procedure, which do not have the force of law (so a court can disregard them) but which the patent examiners follow, so you need to be aware of them;

   all as interpreted by the federal courts.

You can read these things for yourself on the Patent Office website, *http://www.uspto.gov/*.

US006055695A

United States Patent [19]

McKay, Jr.

[11] **Patent Number:** **6,055,695**

[45] **Date of Patent:** **May 2, 2000**

[54] **LINT ROLLER ASSEMBLY**

[75] Inventor: **Nicholas D. McKay, Jr.**, Birmingham, Mich.

[73] Assignee: **Helmac Products Corporation**, Flint, Mich.

[21] Appl. No.: **09/104,605**

[22] Filed: **Jun. 24, 1998**

[51] Int. Cl.[7] **A47L 25/00**

[52] U.S. Cl. **15/104.002**; 15/230.11; 492/13

[58] Field of Search 15/104.002, 230.11; 492/13, 14, 19

[56] **References Cited**

U.S. PATENT DOCUMENTS

3,156,938	11/1964	Bills	15/104.002
3,201,815	8/1965	Selby	15/104.002
3,386,124	6/1968	Feine	15/230.11
4,361,923	12/1982	McKay	15/104.002
4,422,201	12/1983	McKay	15/104.002
4,557,011	12/1985	Sartori	15/104.002
4,570,280	2/1986	Roth	15/104.002

Primary Examiner—Mark Spisich
Attorney, Agent, or Firm—Gifford, Krass, Groh, Sprinkle, Anderson & Citkowski, P.C.

[57] **ABSTRACT**

A lint roller assembly is disclosed for rotatably supporting a tubular and cylindrical adhesive lint remover roller. The assembly includes a pair of elongated housing parts which are substantially identical to each other. Each housing part includes an elongated handle section and a semi-cylindrical lint roller support section longitudinally adjacent the handle section. The lint roller support section has an outer diameter less than the diameter of the adhesive roller. The housing parts are secured together in a facing relationship by registering pins and sockets formed on the housing parts. In doing so, the lint roller support sections form a cylindrical lint roller support for the adhesive roller while the handle sections abut against each other to form a handle.

11 Claims, 3 Drawing Sheets

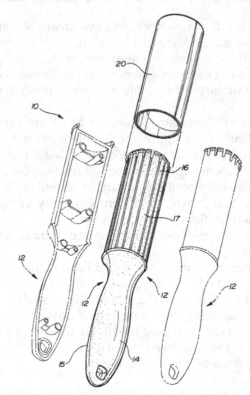

Figure 1.1   United States Patent 6,055,695 "Lint Roller Assembly," cover page.

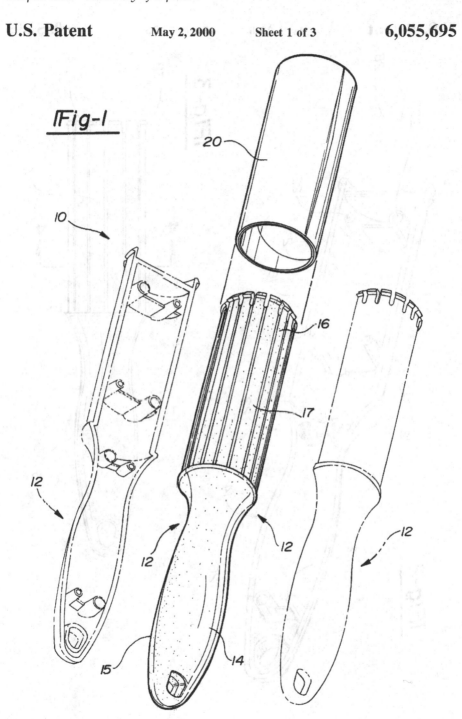

Figure 1.2a   United States Patent 6,055,995, Figure sheet 1 of 3.

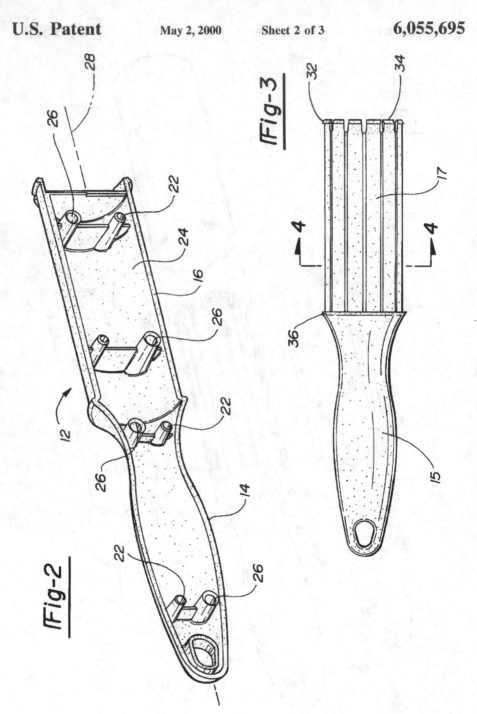

Figure 1.2b   United States Patent 6,055,995, Figure sheet 2 of 3.

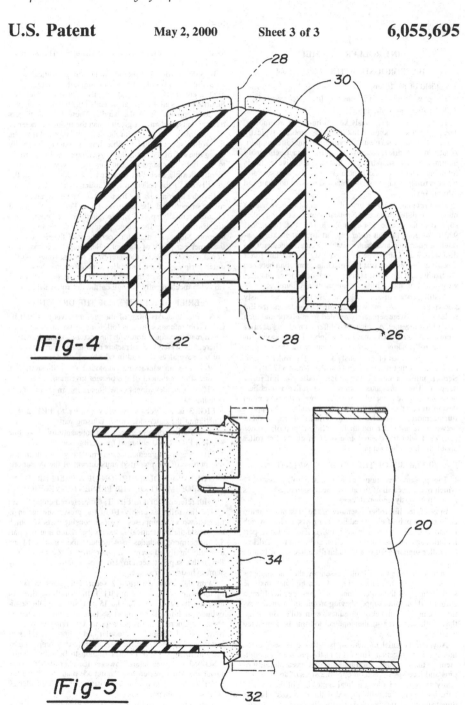

Fig-4

Fig-5

Figure 1.2c   United States Patent 6,055,995, Figure sheet 3 of 3.

6,055,695

1

LINT ROLLER ASSEMBLY

BACKGROUND OF THE INVENTION

1. Field of the Invention

The present invention relates to a lint roller assembly.

2. Description of the Prior

There are many previously known lint roller assemblies. These previously known lint roller assemblies typically comprise a handle secured to a cylindrical lint roller support. A tubular cylindrical adhesive lint roller is then removably mounted to the support such that the adhesive roller is rotatable relative to the handle. In use, the adhesive lint roller is rolled along a user's clothes to remove lint, hair and other debris.

The previously known lint roller assemblies have used a number of different means to rotatably mount the lint roller support to the handle. For example, in U.S. Pat. No. 4,361, 923 to McKay, the lint roller support and handle are separately constructed and then rotatably secured together. One disadvantage of this type of previously known lint roller assembly, however, is that the rotatable connection between the handle and lint roller support is subjected to mechanical wear and tear and, ultimately, mechanical failure.

A still further disadvantage of this type of previously known lint roller assembly is that both the handle and the lint roller support were separately molded from plastic and then assembled together. As such, two different molding cavities were required for each lint roller assembly, i.e. one cavity for the lint roller support and a second cavity for the handle.

Still other types of previously known lint roller assemblies such as that disclosed in U.S. Pat. No. 4,5577,011 to Sartori, utilize a unitary lint roller handle and lint roller support. These previously known lint roller assemblies, however, require a complex and, therefore, expensive mold design in order to mold the lint roller handle and support. Furthermore, a relatively large frictional engagement between the lint roller and the lint roller support oftentimes interfered with the desired free rotation of the lint roller about the lint roller support.

SUMMARY OF THE PRESENT INVENTION

The present invention provides a lint roller assembly which overcomes all of the above-mentioned disadvantages of the previously known devices.

In brief, the lint roller assembly of the present invention comprises a pair of elongated housing parts, each of which is substantially identical to each other. Each housing part includes an elongated handle section and a semi-cylindrical lint roller support section longitudinally adjacent the handle section.

In order to form the lint roller assembly, the housing parts are secured together in a facing relationship such that the semi-cylindrical lint roller support sections together form a cylindrical lint roller support having an outside diameter less than the inside diameter of the adhesive lint roller. Similarly, the handle sections also abut together to form the completed handle.

Any conventional means can be utilized to secure the housing parts together. However, in the preferred embodiment of the invention, each housing part has a plurality of pins and recesses formed on opposite sides of and equidistantly spaced from a longitudinal center line of the housing part. Each pin is laterally aligned with one recess which is dimensioned to receive and frictionally engage its corresponding pin to thereby frictionally secure the housing parts

2

together. Optionally, an adhesive can be utilized between the housing pins and recesses.

In order to rotatably mount the tubular cylindrical adhesive lint roller to the lint roller support, a plurality of circumferentially spaced and radially outwardly extending flexible fingers are formed at the end of the lint roller support sections opposite from the handle. These flexible fingers have an outside diameter greater than the inside diameter of the lint roller such that, with the lint roller inserted over the lint roller support, the flexible fingers extend outwardly along one end of the lint roller thus entrapping the lint roller to the lint roller support. The fingers, however, are sufficiently flexible such that they flex inwardly to permit the installation of the roller onto the roller support.

In order to minimize the frictional contact between the lint roller support and the adhesive lint roller, a plurality of circumferentially spaced and longitudinally extending ribs are formed along the outer periphery of the lint roller support. Since only the outer periphery of the ribs contact the inner periphery of the lint roller, only a minimal frictional contact between the interior of the lint roller and the outer periphery of the lint roller support occurs thus facilitating free rotation of the lint roller about its support. These ribs also mechanically strengthen the lint roller support.

BRIEF DESCRIPTION OF THE DRAWING

A better understanding of the present invention will be had upon reference to the following detailed description when read in conjunction with the accompanying drawings, wherein like reference characters refer to like parts throughout the several views, and in which:

FIG. 1 is an elevational exploded view illustrating a preferred embodiment of the present invention;

FIG. 2 is an elevational view illustrating an inside of one housing part;

FIG. 3 is an elevational view similar to FIG. 2, but illustrating the outer side of the housing part;

FIG. 4 is a sectional view taken substantially along line 4—4 in FIG. 2 and enlarged for clarity; and

FIG. 5 is a fragmentary diagrammatic view illustrating one portion of the preferred embodiment of the invention.

DETAILED DESCRIPTION OF A PREFERRED EMBODIMENT OF THE PRESENT INVENTION

With reference first to FIG. 1, the preferred embodiment of the lint roller assembly 10 of the present invention is thereshown and comprises a pair of housing parts 12 which are substantially identical to each other. Each housing part 12, furthermore, includes an elongated handle section 14 and a longitudinally adjacent lint roller support section 16. The lint roller support section 16, furthermore, is generally semi-cylindrical in shape.

With the housing parts 1:2 secured together in facing relationship as illustrated in FIG. 1, the handle sections 14 abut together to form a handle 15 and, similarly, the semi-cylindrical lint roller support section 16 abut together to form a cylindrical lint roller support 17 having a predetermined outside diameter. A tubular and cylindrical adhesive lint roller 20 is then inserted over and rotatably supported by the lint roller support 17 in a fashion to be subsequently described in greater detail. However, the lint roller 20 has an inside diameter greater than the outside diameter of the lint roller support 17 to permit free rotation of the lint roller 20 relative to the support 17.

With reference now to FIGS. 2 and 4, in order to secure the housing parts 12 together, the housing part 12 includes

---

Figure 1.3a   United States Patent 6,055,995, text columns 1 and 2.

6,055,695

3

at least one pin **22**, and preferably several longitudinally spaced pins **22**, protruding outwardly from an inside surface **24** of the housing part **12**. A corresponding, recess **26** is provided for each pin **22** such that the recess **26** is both laterally aligned and parallel to its corresponding pin **22**. Furthermore, as best shown in FIG. **4**, each pin **22** and recess **26** pair is laterally equidistantly spaced from a longitudinal center line of the lint roller support section **16**. Consequently, when the housing parts **12** are in a facing relationship (FIG. 1) such that each pin **22** registers with its corresponding recess **26**, the outer edges of the lint roller support section **16** meet each other in flush engagement.

In the preferred embodiment of the invention, both the pins **22** and recesses **26** are circular in cross-sectional shape although other shapes may alternatively be used. Furthermore, the outer diameter of the pin **22** is substantially the same, or slightly greater than, the inside diameter of its corresponding recess **26** so that, with the housing parts **12** pressed together, the frictional engagement between the pins **22** and their corresponding recesses **26** frictionally lock the housing parts **12** together. Optionally, an adhesive can be used between the pins **22** and recesses **26**.

With reference now to FIGS. 3 and 4, a plurality of circumferentially spaced and longitudinally extending ribs **30** are formed along the outer surface of the lint roller support **17**. These ribs **34** operate not only to mechanically strengthen the lint roller support **17**, but also to minimize the frictional engagement between the outer periphery of the lint roller support **17** and the inner periphery of the lint roller **20**.

With reference now to FIG. 5, a plurality of circumferentially spaced and radially outwardly extending flexible fingers **32** are formed around an end **34** of the lint roller support **17** most distant from the handle **15**. In their unflexed state, the outer diameter of these fingers **32** is greater than the inside diameter of the lint roller **20**. However, by flexing inwardly, the fingers **32** permit the lint roller **20** to be inserted over the end **34** and onto the lint roller support **17**. A radially outwardly extending flange **36** at the junction between the handle **15** and lint roller support **17** has a diameter greater than the inside diameter of the lint roller **20** such that the lint roller **20** is longitudinally trapped on the lint roller support **17** between the flange **36** and the fingers **32**.

From the foregoing, it can be seen that the lint roller assembly of the present invention provides an economical yet durable lint roller assembly. Furthermore, since the housing parts are substantially identical to each other, only a single mold cavity design is required to form the entire lint roller handle and support.

Having described my invention, however, many modifications thereto will become apparent to those skilled in the

4

art without deviation from the spirit of the invention as defined by the scope of the appended claims.

I claim:

1. A lint roller assembly comprising:
   a tubular and cylindrical adhesive lint roller,
   a pair of elongated housing parts, each housing part being substantially identical to the other,
   each housing part having an elongated handle section and a semicylindrical lint roller support section longitudinally adjacent the handle section, said semicylindrical lint roller support section having an outer diameter less than the diameter of the roller,
   means for securing said housing parts together in facing relationship such that said semicylindrical lint roller support sections together form a cylindrical lint roller support and said handle sections abut against each other to form a handle.

2. The invention as defined in claim 1 wherein each housing part is of a one piece construction.

3. The invention as defined in claim 2 wherein each said housing part is made of a plastic material.

4. The invention as defined in claim 1 wherein each semicylindrical support section includes a plurality of circumferentially spaced and radially outwardly extending flexible fingers around its end most distant from its associated handle section, said fingers having an outer diameter greater than an inner diameter of the roller.

5. The invention as defined in claim 1 wherein each lint roller support section includes a plurality of axially extending and circumferentially spaced ribs along its outer periphery.

6. The invention as defined in claim 1 wherein each housing part is of a one piece molded plastic construction.

7. The invention as defined in claim 6 wherein said handle sections and said support sections are coaxially aligned.

8. The invention as defined in claim 1 wherein said securing means comprises at least one pin and one recess, said pin and said recess being formed on opposite sides of and equidistantly spaced from a longitudinal centerline of each said housing part, said recess dimensioned to receive said pin, and said pin and said recess being laterally aligned with each other.

9. The invention as defined in claim 8 and comprising a plurality of longitudinally spaced complementary pins and recesses.

10. The invention as defined in claim 8 wherein said recess is dimensioned to frictionally engage said pin upon insertion.

11. The invention as defined in claim 10 wherein said pin and said recess are circular in cross-sectional shape.

* * * * *

Figure 1.3b United States Patent 6,055,995, text columns 3 and 4.

1.1.1.1 *The right to exclude, not to do*

A patent is a right to exclude, not a right to do. You may understand those words, but invariably there will come a time when you do not believe them. Alas, they are not debatable.

Here is an example. Imagine that I invent and patent the wheel. You invent and patent the bicycle. Which of us can manufacture bicycles? Answer: nobody. I cannot make bicycles, because your patent excludes everyone from making bicycles unless you authorize them to do so. (Note: The ordinary way to authorize is to license. See Chapter 10.) But what about you? Unless you are happy to sell wheel-less frames, you cannot sell bicycles either. I have the patent on the wheel, and I can exclude you from making wheels. What happens?

1. I can *sell* you wheels that you incorporate in your bicycles. Now we are both happy. (That is because if you buy something from the patent owners, or someone they have licensed, you are automatically licensed to use it, unless by an explicit term in a contract they have placed some restrictions on your activities are restricted.)
2. You can get a license from me to make your own wheels.
3. We can go to court, and either (a) a judge will decide who is right (right about *what*? you ask. We will get to the basic issues of patent law later) or (b) we will settle the lawsuit among ourselves. Most patent litigation ends in settlement, so if anyone does go to court, the case will probably be settled. The settlement may take the form of me buying your company, or you buying mine, or maybe a cross-licensing deal. Or perhaps one of us will sell or assign our patent rights to the other in exchange for money, and go do something else.

We will revisit **the right to exclude but not to do** again, when we talk about patents as prior art (see Section 1.2.2). "Prior art" is another term of art in patent law. The meaning may be exactly what you would guess, but it is good to be wary about throwing around terms of art unless you are quite confident you are using them correctly. The wrong term of art in the wrong place could lead to confusion, or worse, liability. "Prior art" refers to anything in your field of technology before your invention. When does "before" end? That is another complicated question, which is discussed in later chapters.

1.1.1.2 *Patents have nationality*

Patents are grants from the sovereign (the government). This means that they have nationality. My U.S. patent gives me rights in Michigan, but if I cross the Ambassador Bridge over to Ontario, my patent is useless. If I want the same kind of rights on the other side of the bridge, I must apply for a patent in Canada. Anyone of any nationality may apply for a patent in any country, but each country can only issue a patent that confers rights in that country. (Note: The European Union has considered having a pan-European patent but in November 2002 rejected the idea yet again. By the time you read this book, things may have changed. One place to look for information is the website for the European Patent Office, *http://www. european-patent-office.org*.)

Worldwide patent protection is costly, as you might imagine. The good news is that you can start the process for multicountry patenting and then have some time to decide whether your invention is really going to be commercially successful, and where you are most likely to want to have a patent. This is called "filing a PCT application" where PCT stands for Patent Cooperation Treaty.

1.1.1.3   Patents v. copyrights, patents v. trademarks

If you are reading this book, chances are that you know the difference between patents, copyrights, and trademarks. But just to review, some comparisons are shown in Table 1.1.

Might a typical reader of this book want all three kinds of protection? Sure. Say you invent a new device *and* you write a manual or instructions for use *and* you make up a clever name and logo for the device or for the company that will make or market it. You will want to consider patent protection for the invention, copyright protection for the instructions (they are not likely to be all that "creative," and the copyright will be a "thin" one, but at least you would have another arrow in your quiver if someone sells knock-offs of your invention and simply photocopies your manual), and trademark protection for your name and logo. (And you will probably want to register a domain name for use on the Internet, which is separate from obtaining trademark protection and is even more remote from the subject of this book.)

1.1.1.4   Patenting v. trade secrets

1.1.1.4.1   Differences.   Instead of going to the expense in time and money of obtaining a patent for your invention, you might want to keep it as a "trade secret." Or you might want to patent the basic invention, but later, as you develop a more cost-effective or less-polluting or faster process for making it, you might want to keep those improvements to the commercial process as trade secrets. What is the difference beween a patent and a trade secret?

Length of time of protection.   A **trade secret** is protectable forever — as long as it is kept secret. If the secret is not discoverable by independent investigation (i.e., some way other than stealing it from the secret-holder), and is not disclosed by the holder, then it remains a protectable secret. Trade secrets can be licensed to people who promise to keep the secret, and thieves can be sued for misappropriation.

Table 1.1 Comparison between Patent, Copyright, and Trademark Features

Questions	Patent	Copyright	Trademark
What is protected?	Inventions that are new and useful[a]	Creative expression	Symbols, words, or phrases that designate the origin (manufacturer, producer) of goods and services
The words in the top row are not quotations from the statutes. If I want to see the statutes, where do I start?[b]	35 USC 101	17 USC 102	15 USC 1051
What verb describes what I do to get a [patent, copyright, or trademark]?	Apply	Register	Apply (then, if successful, register)
To whom or with whom do I [that verb]?	United States Patent and Trademark Office	Library of Congress, Copyright Office	United States Patent and Trademark Office
Patents, copyrights, and trademarks are creations of federal (national) law. Is there parallel or similar protection under state law?	No	No	Yes

[a]These are "utility" patents. You can also patent a "design" and a "plant" (see Section 1.3.1) but when most people think of patents, they are thinking of *utility* patents.

[b] One place to see the text of statutes is *www.findlaw.com/casecode/uscodes*.

A **patent**, on the other hand, has a term set by statute. The patent is not in force (you cannot sue anyone for infringement) until it is issued by the Patent Office. It expires 20 years after the earliest filing date, which is usually at least 2 years before it was issued. There are exceptions at each end, but this statement is a good general rule to start with: a patent has a life of about 17 or 18 years.

Costs.   The costs associated with **trade secret** protection vary, depending on how you want to keep the secret, and how you make sure it stays secret. You may hire counsel to draft nondisclosure agreements, you may hire security companies to provide human or electronic security or you may just buy locks.

By contrast, obtaining a **U.S. patent** has somewhat predictable costs. From application through issuance, a patent is likely to cost in the neighborhood of $20,000 or more. There are fees to the Patent Office for applying and for issuance (as well as for missing deadlines, for getting expedited examination, for maintaining the patent in force every 4 years after issuance until expiration, and so on). The Patent Office current fee schedule can be found at *http://www.uspto.gov/go/fees/fees.htm.*

In addition to paying Patent Office fees, most inventors use the services of a patent attorney or patent agent (someone licensed to practice before the Patent Office). These fees vary depending on the complexity of the invention, the technology generally, the existence of other patents and applicants and the degree of similarity with their inventions, the way things unfold at the Patent Office (many aspects of which may not be controllable by the applicants or their patent attorneys or agents), etc.

Damages (what a court can order an infringer to pay you). The **patent** statute and the case law that has interpreted it are generous to patent owners. Your "damages" (in legal parlance, the word "damages" refers to the money you win), can be no less than "a reasonable royalty" — the royalty the infringer should have paid you to begin with. In addition, you may receive "lost profits," which could include amounts you would have made on the patented item but for the infringement, as well as amounts from "convoyed sales" — other aspects of your business that might have done better if the infringer had not been in the market. If your case is deemed "exceptional," you can also receive reimbursement of your attorney fees (which in patent litigation can run into millions of dollars because the litigation can take many years to resolve) and the damages can be multiplied up to three times.

Trade secret damages are governed by state law. Most states have enacted the Uniform Trade Secrets Act, and it provides for compensatory damages, and damages for unjust enrichment, for the misappropriation. If the misappropriator acted willfully and maliciously, then the damages can be doubled. The Uniform Trade Secrets Act follows the usual rule that each side pays its own attorney fees, but there are exceptions such as if the claim of misappropriation by the trade secret owner was made in bad faith, or if the misappropriation was willful and malicious.

1.1.1.4.2 Why choose to patent? Given the shorter term and the likely higher costs, why would you ever choose to patent? The obvious reason is that it is hard to keep a secret. Also, the upside potential from a lawsuit may be worth something, not just in its own right, but also as leverage in bargaining to license your invention. In addition, other people you deal with or would like to deal with (investors, lenders, and even key employees) may also want you to have patents among your assets and may expect you to be patent-savvy.

1.2   Parts of a patent, and how to read one

1.2.1   Specification, figures, and claims

A patent, such as the one shown in Figure 1 on pages 6 to 12, has a cover sheet (Figure 1.1) and, after that, three major parts: the figures (Fig. 1.2a through c), the specification (Figure 1.3a and 1.3b up to column 4, line 2), and the claims (Figure 1.3b, column 4, line 3 to line 50).

The specification and the claims are written in words, and every patent must have both. The figures are pictures that show how the invention is put together and how it works. All patents except chemical composition patents must have figures. Chemical composition patents generally have formulae. Each figure must be numbered, and anything on a figure — whether it is a whole area, a specific piece, a feature on a specific piece, or a hole or void — must be individually numbered if it is discussed in the specification. For example, in Figure 1.2a and Figure 1.2b, the recess designated with numeral 26 appears in the patent drawings and is discussed in column 3 at various places, such as lines 3 to 8 (Figure 1.3b).

The specification is the place for all the "talk." It must include an explanation of the figures. It must give the educated reader sufficient information to make and use the invention. This is called the **enablement** requirement. The concept in patent law of the educated reader is known as the **person of ordinary skill in the art**. The specification must also state the inventor's preferred way to make and use the *claimed* invention (but need not state preferences as to things that are not *claimed*). This is the **best mode** requirement. All the terms in boldface are discussed in greater detail elsewhere in this book.

The claims are the words that define your invention **legally**. Patent lawyers like to say they define the "metes and bounds" of the invention, just as real estate on a deed is defined by its "metes and bounds." Whenever you consider a patent, whether to make sure it is valid or to see if someone infringes it or should be offered a license, you must look to the claims. "Construing the claims" (deciding what the words means) is often difficult mental work; drafting good claims can be even harder.

1.2.2   Patent as instrument of legal rights v. patent as "prior art"

To understand the differences between a patent as a declaration of legal rights, on the one hand, and a patent as "prior art," on the other, you will need to understand the phrase **read on** and be comfortable with the use of a **claim chart**.

If we are interested in whether or not your claims are valid, we must compare them to the prior art. But the prior art undoubtedly includes other people's patents. What part of *those* patents is most important if you want to know if *your* patent is *valid*? What part of *those* patents is most important if you want to know if you can practice *your* invention without *infringing* those patents? The answers to those questions are different. Your patent is valid if *your* claims do not **read on** the prior art. You will not infringe those other patents if *their* claims do not **read on** the object you make, use, sell, offer for sale, or import, or, if the patent is on a process or method, the processes or methods you use.

1.2.2.1 "Read on"

The grammatical subject of "read on" is always a *claim*. The grammatical object of "read on" is either

- an accused device, if the question is whether or not the claim is *infringed*, or
- the prior art, if the question is whether or not the claim is *invalid*.

Thus we may say that claim 5 of the patent **reads on** the competitor's model 23-BQ (so model 23-BQ infringes that claim). Or claim 5 may **read on** the competitor's old model 15-LJ, which was sold 50 years ago throughout the U.S., in which case claim 5 is invalid.

In order for a patent owner to prevail against an accused infringer, the claim in question must be both valid and infringed. (In addition, the patent as a whole must be free of inequitable conduct, discussed in Chapter). Thus the infringement inquiry focuses on the **claim** of the patent in question.

But when a patent is relevant not for the owner's rights but as prior art to someone else's patent or patent application, then we care about what it **teaches**. We care much less about what it claims. (Generally the claims do not teach anything that is not taught elsewhere in the patent; if they do, then they may be invalid.) The primary reason you look at the claims of a prior art patent is to make sure you will not be infringing it, or to see a way to design around it. But **infringement** of a prior art patent by practicing your patent is a different inquiry from **invalidation** of your patent by that prior art patent.

1.2.2.2 Claim charts

A claim chart is a two-column table in which you compare the language of a patent claim to whatever it is you care about — a possibly infringing device or process, or the prior art. In the left hand column you put the claim, broken down into its conceptual pieces[*] to suit the inquiry of the claim chart maker:

* Patent lawyers may speak of the "elements" of the claim. That can be confusing. Sometimes "element" may refer to a labeled clause, when the claim is written in outline form (1 – A – (i), etc.). Such a claim might have elements 1 through 5, and element 4 might have sub-elements A and B, etc. But it may be that for the purposes of a claim chart, elements 1 through 3 can be lumped together because everyone would agree that they are found in whatever the claim is being compared to. Element 5, however, may need to be broken down into its individual words. I use the colloquial "pieces" to emphasize that the division of the claim is ad hoc.

For example, in the printed patent some parts of the claim may be indented and/or labeled with letters and numbers. (In Figure 1.3b, claim 1 is format-ted with indentation, at column 4, lines 4 to 17.) If some items are not controversial, the claim chart maker may put them together into a single row of the chart. Alternatively, a three-word phrase in the claim may be the essence of the conflict, and each word's scope is contested. In that case, each of the three words might occupy its own row. There are no rules; logic and intelligence dictate how you make the chart.

Whether you are considering infringement or validity, look for a 1:1 correspondence between the parts of the claim and whatever it is being compared to. If there is a 1:1 correspondence between the claim and the accused device, the claim is infringed. If there is a 1:1 correspondence between the claim and the prior art, the claim is invalid.

Consider the Lint Roller Assembly Patent's claim 1 (Figure 1.3b, column 4, lines 4 to 17). Compare it to lint roller assemblies you may have used long before the patent was applied for. The application date is listed on Figure 1.1, on the right side of the patent cover sheet. This patent application was first filed on June 24, 1998. You would probably guess that every lint roller since lint rollers were first made would have "a tubular and cylindrical lint roller" (the first indented phrase) and a roller support that has a smaller diameter than the roller itself (the third indented phrase). But the second indented phrase, which specifies that the housing is made from two "sub-stantially identical parts," is different. The words in that part of the claim would have to be considered very carefully, whether the question was infringement or invalidity.

1.2.2.3   A useful table about specification and claims

Table 1.2 reinforces what you should know already. Let us say you have just obtained a patent. Let us call it the New patent. Your competitor has a patent that is older than yours, the Old patent. You may be concerned about whether your New patent is valid over the Old patent, and you may be concerned about whether, in practicing your own invention (now covered by the New patent), you need a license under the Old patent.

Wait, you may say: if the Patent Examiner granted me the New patent when she knew all about the Old patent and maybe even initially rejected my claims in light of it, doesn't that mean I could not possibly infringe the Old? The answer is a resounding *no*. If you find this puzzling, you are in excellent company with some of my best law students, chief executive officers (CEOs), federal judges, etc. But I hope that you will not be puzzled after you have had a chance to study Table 1.2, and to think about patents, their specifications — where they *teach* — and their claims — where they wall off the owner's area of exclusivity.

The reason that the answer is *no* is that validity is a different inquiry from infringement. Validity of the New patent depends on the New patent's claims. Infringement of the Old patent depends on the Old patent's claims.

1.3 The application for a patent: what kind of patent? what kind of application?

1.3.1 Kinds of patents

The Patent Office issues three kinds of patents: utility patents, design patents, and plant patents. If you are reading this book, chances are you want a **utility** patent, a patent on an invention. The claims in a utility patent address how a device or process is made or works. They may even address how the object is "designed" in the usual sense of the word "designed" (on a drafting board, with rulers and templates, or on a computer using a CAD program) so that the claimed invention functions in the way the inventor had in mind. But the word **design** in the phrase "design patent" is another term of art.

Table 1.2 Infringement and Validity Assessment between Old Patent and New Patent

Roles of New and Old	"New" is the patent of interest. "Old" is the Prior Art.	"Old" is the patent of interest. "New" describes the accused device, because it is made according to the New patent.
	*Is the New patent **valid** over the Old patent?*	*Is the Old patent **infringed** by someone practicing the New patent?*
New patent	Look at New patent's **CLAIMS**	Look at New patent's **SPECIFICATION** to understand what someone practicing the New patent would do.
Old patent	Look at Old patent's **SPECIFICATION** to see what it "teaches."	Look at Old patent's **CLAIMS**.

A **design** patent is available only to protect the way your device looks, and then only to the extent that how it looks is **purely ornamental** and **not at all** functional. Thus the design of indentations on the bottom of a running shoe could be protectable by a design patent, but if the shoe advertised that its soles were unusually good at gripping wet surfaces, the design patent could be invalid.

To put it another way: just because you have "designed" something does not mean you want a *design* patent. In fact, usually you do not. A "design" patent protects only the appearance of the object, and the elements of that appearance that are dictated purely by esthetic considerations. If the appearance of an item has a functional purpose, then it is not patentable as a *design*. The item may, however, be entitled to a *utility* patent.

A *plant* patent is for a new variety of plant that is reproduced asexually. Plants that reproduce by seeds are not patentable, but they may be entitled to certificates from the Department of Agriculture under the Plant Variety Protection Act.

1.3.2   Kinds of applications, and a non-application

Until the mid-1990s, if you wanted to apply for a patent, you had to have a complete draft to submit to the Patent Office, including claims. Then U.S. law changed, to make it more in harmony with the laws of other countries. As part of those changes, the "provisional" application came into existence. A provisional application does not need claims, and has the advantage of giving the applicant a filing date for whatever disclosure is made in the provisional. Thus researchers who want to give a talk at a conference about their work, or want to submit an article for publication in a prestigious journal, may file their talk or manuscript as their provisional application and safeguard themselves against having created invalidating prior art. (For a discussion of why the talk or manuscript might be invalidating prior art, see Chapter 3.) Similarly, companies wanting to show a device — or even a prototype or a subassembly — at a trade show can file their literature and diagrams as the provisional. This will give them an advantage, especially if they want to file in countries (such as Europe and Japan) where there is no grace period for an inventor to publish, sell, or offer to sell an invention before filing the patent application.

1.3.2.1   Provisional

The provisional application need not have claims or an inventor oath, and is much cheaper to file than a regular application. An independent inventor or anyone entitled to reduced rates (university researchers, small businesses) is currently charged a fee of $80. The cost of a regular application, assuming it does not have more than the basic numbers of independent and total claims, is $395. (The fees are double if the inventors have assigned their rights to larger companies.) Filing a provisional gets the filer a date, but does not obligate the Patent Office to do anything with the application. The Patent Office's work begins only when the regular application is filed and the regular fee paid.

1.3.2.2 *Regular application*
A regular application, as indicated above, costs more, must include at least one claim, and triggers the process of examination by the Patent Office.

1.3.2.3 *Statutory invention registration*
Occasionally inventors may want their invention to be part of the patent literature, so that their disclosure is searchable and findable at the Patent Office. They do not, however, want to obtain a patent. In that case they may file a Statutory Invention Registration. This makes their invention part of the "prior art" as of its filing date. It therefore has defensive value, but no offensive value (in the monetary and litigation sense).

1.4 *Requirements of a utility patent*

In addition to a utility patent having the parts specified by the statute — specification, figures, and claims — it must have certain content. The statute (35 USC section 101) says that "whoever invents [something] **new** and **useful**...may obtain a patent...."

What is it that must be new and useful? Answer: the invention described in each claim, and each claim is evaluated separately. So when I refer to the "claimed inventions" with an "s" at the end, I am emphasizing that the invention in claim 1 can be different from the invention in claim 12 (in ways that are crucial to the proof that the claim is valid or that it is infringed). I do not mean to say that you have more than one patent, but only that your one patent has claims of various scope.

How do you tell the public that your claimed inventions are **useful**? Answer: you can explain the utility in the specification if it is not instantly clear, and you draft the claims to make sure that the usefulness will be garnered from anyone practicing the invention. How do you tell the public that your invention is **new**? Answer: you can hype, tout, and otherwise praise your invention in the specification, and describe how it overcomes problems of the prior art, but the true test will be whether your claims have been drafted so that they do not **read on** the prior art.

This is just another reminder that the specification and claims are different animals, that their roles are different, and that you have to pay attention to both, whether you are applying for a patent or scrutinizing someone else's.

1.4.1   The invention — especially *as* claimed

1.4.1.1   Utility

Most inventions have obvious utility, and the utility requirement is not a source of controversy. There are a few exceptions, however. One is for unbelievable inventions. The classic example is a perpetual motion machine. People used to say that cures for the common cold were also inherently suspect, but with the biotech revolution maybe that is no longer true. Claims to perpetual motion machines may be challenged on the grounds of utility, but they also may be challenged on **enablement** grounds (see Section 1.4.2.2). The other class of inventions that have recently been subject to utility challenges are in the biotechnology area. They may be intermediate biological materials whose use is unknown, or whose only use is as a possible means to something else, as yet unknown or undoable. If your invention is in the biotech area, you will need to investigate the current requirements for utility. The best place to start will probably be at the Patent Office website, *http://www.uspto.gov/web/offices/pac/dapp/utility.htm*, where there is a document concerning the utility requirements.

1.4.1.2   Novelty

"Novelty" in patent law means that the claim does not **read on** the prior art. Whether a claim is in an application and a Patent Examiner has to decide if it is "new," or the claim is in an issued patent and a potential licensee, or litigant, wants to decide if it is valid with regard to the novelty requirement, the person evaluating the claim should at least mentally, if not with pencil and paper (or computer), make a claim chart. Every element, or better, piece (see Section 1.2.2.2) of the claimed invention must be found in the prior art in order for the claim to be invalidated as *not* new. But there are two ways to show lack of novelty, and they have several different aspects. One way is to find a **single** piece of prior art — a device that has already been made or sold, a patent that teaches about every aspect of the claimed invention, or a journal article, advertisement, or manual that completely describes every feature that is mentioned in the claim. If a **single** piece of prior art invalidates the claim, the claim is said to be **anticipated**.

The other way to show lack of novelty is to use more than one piece of prior art, or one piece of prior art that almost, but not quite, teaches every aspect of the claimed invention, but which, when coupled with the knowledge of "the person of ordinary skill in the art," has it all. Then the claim is said to be **obvious** in light of the prior art.

Anticipation has the advantage of simplicity (only one piece of prior art need be considered) for both the patent owner and the patent challenger. But an attack based on obviousness is harder for the challenger for other reasons besides complexity. We address that in the next section.

1.4.1.3 Nonobviousness

If there is no single piece of prior art that anticipates a claim, the claim may nonetheless *not* be new. It may be "obvious to a person of ordinary skill in the art." This means that more than one piece of prior art, or one or more pieces plus the knowledge of that person (sometimes called "the ordinary artisan," and sometimes called by the acronym "POSITA") invalidates it.

As soon as the attack goes from anticipation to obviousness, the attacker has some other problems: first, the attacker must show that there is a "motivation or suggestion to combine" the pieces of prior art in the prior art itself; and second, the patent defender can invoke "secondary considerations" to demonstrate that what may look like obviousness really is not. Secondary considerations are practical events which provide circumstantial evidence that the invention is new relative to the prior art. Some are post-invention events that suggest that the world thinks well of the invention. They include commercial success of the inventor (or even of copiers) and recognition by others (such as being named inventor of the year, or having journals and text books salute the invention). Others are events that predate the invention, such as the failure of others to reach a solution that the inventor has achieved, and the long-felt need for the invention.

1.4.2 The invention, especially what is in the specification

The Patent Office has information on the requirements for the specification at *http://www.uspto.gov/web/offices/pac/dapp/35usc112.htm*. This web page is entitled "35 U.S.C. 112 Rejections Not Based On Prior Art." There are three requirements for the specification in the statute, specifically in Section 112, paragraph 1. They are called the **written description** requirement, **enablement**, and **best mode**. The statute, formatted to show the three requirements (and to show why until recently they were thought of by many people as being only two — enablement and best mode) says:

> [¶1] The specification
> shall contain a written description
> > of the **invention**, and
> > of the manner and process of making and using it, in such full, clear, concise, and exact terms as to **enable** any person skilled in the art to which it pertains, or with which it is most nearly connected, to make and use the same,
> and shall set forth the **best mode** contemplated by the inventor
> of carrying out his invention.

Although the requirements of Section 112, paragraph 1 are phrased in terms of what the specification says, remember that validity always centers on a particular claim: it is the **claimed** invention that must be adequately described in the specification, that the specification must enable the POSITA to make and use, and for which the inventors must describe their best mode of making and using as of the date of their application. Often the claimed invention and the general idea described in the specification are the same, and the specific words of the claim can be safely ignored. Sometimes, however, the claimed invention is more specific, and has a detail that is not central to the general idea of the invention. In that situation, attackers of the patent on the grounds of Section 112 must once again think in terms of a claim chart so that they can show how the specification fails to support the **claimed** invention.

Generally, attacks under the best mode and enablement provisions of Section 112 are made in the courts, against issued patents, rather than in the Patent Office, against applications for patent. Among other reasons, this is because the patent examiners may not have the practical knowledge to understand whether the POSITA is enabled to make and use the invention, nor the inside knowledge to know whether the inventors have disclosed their best mode.

1.4.2.1 Written description

The written description requirement means that the claim(s) must bear out what the specification indicates is a central or essential feature of the invention. Thus if a specification says that the whole point of the invention is to use fried eggs, but the claims just say eggs, then the written description and the claims do not agree. For almost all patents, a "written description" problem emerges, if ever, in litigation. If patent examiners think that the claims and the specification differ, they are more likely to use a different statutory provision, Section 112, paragraph 2, which requires that "The specification shall conclude with one or more claims **particularly pointing out and distinctly claiming** the subject matter which the applicant **regards as his invention**." Rejections of applications claims on the grounds of Section 112, paragraph 2 are fairly common. The examiner will say that there is no "antecedent basis" for a term used in the claim. That is, the term appears for the first time in the claim and was not previously used in the specification. Such rejections can generally be overcome easily, as long as the term was in the claims as originally filed.

1.4.2.2 Enablement

As mentioned above, the enablement requirement means that the inventor must enable POSITAs to make and use the invention. Sometimes an enablement attack might turn on the level of education and experience someone must have in order to qualify as a "person of ordinary skill in the art," but more often the attack turns on whether or not that POSITA could make or use the invention "without undue experimentation." If there are only hints in the specification, and the POSITA would then have to work out problems in the lab over several person-months or even person-years, the patent claims may be vulnerable as not "enabled."

1.4.2.3 Best mode

There are two things to remember about the best mode requirement: first, it is the best mode in the minds of the inventors, not their assistants or bosses or people in the next lab; and, second, it is the best mode at the time the application is filed. Thus if research is kept separate from development, the ideas of the development people do not need to be in the specification (unless they are co-inventors — see Chapter 6).

Timing remains key of course. If inventors disclose their idea to their management, and management hires a patent lawyer, and the patent lawyer works on the application for a few months, then by the date the application is filed the inventors may have a different "best mode" than they had when they started. This means that everyone should have the best mode requirement in mind throughout the application process.

1.5 Case studies

1.5.1 A sample patent: 6,055,695, Lint Roller Assembly (Figure 1, pages 6 through 11)

Look over the specification, including the figures, and the claims of U.S. Patent No. 6,055,695, Lint Roller Assembly. Find the antecedent basis for the word "housing." Where is it discussed? What language *in the claims* is related to the improvement over the prior art concerning the fact that prior art assemblies needed "two different molding cavities" (Figure 1.3a, column 1, lines 28 to 31)? Ask yourself if you could achieve the same economy and simplicity of manufacture as the inventor did, without infringing one or more of these claims? Well-drafted claims will make it hard to design around the patent without sacrificing the inventive advantages for which patent protection was sought.

1.5.2 Should I patent my invention or hold it as a trade secret?

The answer to the question "Should I patent my invention or hold it as a trade secret?" may be "Both." There are some aspects of an invention that may be better kept as trade secrets, and others that should be patented. Quite often, people choose to patent the basic idea (of course, enabling POSITAs to make and use it, at least in the laboratory or on a by-hand basis). Later, when they perfect the "black arts" required to scale it up to commercial manufacture, those techniques are kept as trade secrets.

The enablement and best mode requirements can present problems related to the interplay of patents and trade secrets. Sometimes the patent specification will explain that something must be done with software. If the patent owner has written that code, but is holding it as a trade secret, that could create enablement problems down the road. The question will be how hard it would be for POSITAs to write their own code.

At other times the patent owner will apply for the patent before determining how to scale the invention up to commercial manufacture. After the patent application is filed, the development work will begin. That might create both enablement and best mode problems, but if the claimed invention covers the device itself, and nothing in the claim addresses large-scale production, then the inventors probably do not have to worry about failing to enable anyone to make the invention commercially. As always, the focus must be on what is claimed, not on the general idea discussed in the specification.

Appendix

From *http://www.uspto.gov/web/offices/pac/doc/general/index.html#patent* (last modified 5/1/03):

> Utility patents may be granted to anyone who invents or discovers any new and useful process, machine, article of manufacture, or compositions of matters, or any new useful improvement thereof;
>
> Design patents may be granted to anyone who invents a new, original, and ornamental design for an article of manufacture; and
>
> Plant patents may be granted to anyone who invents or discovers and asexually reproduces any distinct and new variety of plants.

From *http://www.uspto.gov/web/offices/pac/doc/general/index.html#prov:*

> **Provisional Application for a Patent**
> Since June 8, 1995, the USPTO has offered inventors the option of filing a provisional application for patent which was designed to provide a lower cost first patent filing in the United States and to give U.S. applicants parity with foreign applicants. Claims and oath or declaration are NOT required for a provisional application. Provisional application provides the means to establish an early effective filing date in a patent application and permits the term "Patent Pending" to be applied in connection with the invention. Provisional applications may not be filed for design inventions.
>
> The filing date of a provisional application is the date on which a written description of the invention, drawings if necessary, and the name of the inventor(s) are received in the USPTO. To be complete, a provisional application must also include the filing fee, and a cover sheet specifying that the application is a provisional application for patent. The applicant would then have up to 12 months to file a non-provisional application for patent as

described above. The claimed subject matter in the later filed non-provisional application is entitled to the benefit of the filing date of the provisional application if it has support in the provisional application. If a provisional application is not filed in English, then any non-provisional application claiming priority to the provisional application must have a translation of the provisional application filed therein. See title 37, Code of Federal Regulations, Section 1.78(a)(5).

Provisional applications are NOT examined on their merits. A provisional application will become abandoned by the operation of law 12 months from its filing date. The 12-month pendency for a provisional application is not counted toward the 20-year term of a patent granted on a subsequently filed non-provisional application which relies on the filing date of the provisional application.

A surcharge is required for filing the basic filing fee or the cover sheet on a date later than the filing of the provisional application.

A brochure on Provisional Application for Patent is available by calling the USPTO General Information Services at 1-800-786-9199 or 703-308-4357 or by accessing USPTO's website at *http:// www.uspto.gov/*.

Part II

Inventive activities

chapter two

Research records in the patent process

Avery N. Goldstein
Gifford, Krass, Groh, Sprinkle, Anderson & Citkowski, P.C.

Contents

2.1 Introduction

The notebook is a ubiquitous feature in the laboratory setting. While no one can dispute the necessity of documenting experimental work, the effort required to prepare, evidence, and store laboratory data is a frequent lament of the overtaxed researcher and a considerable expense to the organization as a whole. This chapter seeks to highlight the requirements of effective experimental documentation as it relates to the patent process and to provide a logical rationale for documentation procedures.

The legal purpose for documenting experimental work is to provide a trier of fact with a convincing evidentiary paper trial to support the factual assertions of the researcher regarding the date and substance of the invention. A laboratory notebook and a system of documentation add credibility beyond oral evidence and erratic written notes. The goal of documentation is to survive challenges under the rules of evidence and provide a convincing piece of evidence in support of the researcher's legal position. In the U.S., the Federal Rules of Evidence most often need to be satisfied since patent law is a matter under federal, as opposed to state, jurisdiction. All aspects of record-keeping put in place should serve to counter an allegation of forged or falsified laboratory documents. Before elaborating on specific procedures it is worthwhile describing some general principles of good laboratory documentation.

A laboratory documentation system that leaves control over records in the hands of the researcher is vulnerable to challenges of misconduct. To place this statement in context, when laboratory notebooks are used in a legal proceeding to secure or defend patent rights, there is invariably an opposing party with an interest in discrediting the evidence found in the laboratory notebooks. It is not an overstatement that the survival of a corporation as a viable entity may be at stake when decisions about who is the rightful inventor or a patent infringer are involved. An opposing party with so much at stake can be expected to raise every viable challenge to discount the weight a trier of fact will give to various pieces of supporting evidence. This legal strategy can be successful even if the dispute is never brought in front of a trier of fact since it weighs towards a settlement more favorable to the opposing party.

2.2   Legal situations where laboratory records are reviewed

There is a variety of circumstances where laboratory records are considered. The weight given to laboratory documentary evidence should not be under-estimated since it is collected at a time before dispute or contest and, as such, is considered less likely to be colored by personal or financial interests that later develop. In some proceedings before the United States Patent Office, where there is no provision for hearing or evaluating anything other than written evidence, laboratory documentation is often the pivotal evidence. In detailing the varied situations where laboratory records are important pieces of evidence, it is hoped that the reader gains an added appreciation as to the value of implementing a credible record-keeping system.

2.2.1   Predating prior art

In some patent jurisdictions, most notably the U.S., there is a mechanism to "swear behind" a reference that is cited as prior art to reject pending patent claims. When the reference is published up to one year before the filing date of patent application and after the patent applicant's actual date of "redu-cing the invention to practice," one is allowed to provide evidence as to the timing of invention relative to the reference publication date. In this context, reduction to practice means more than having the conception of the inven-tion and includes getting the invention to work in some form. The details of this procedure are found in 37 CFR §1.131. This procedure usually involves a declaration prepared by a patent professional to incorporate the state-ments of the inventor or other knowledgeable individual regarding the timeline of the invention. Documentary evidence is appended to the declar-ation in support of the statements made by the declarant.

2.2.2   Interference

Laboratory research documents are also needed to establish inventorship in those jurisdictions that reward the first to invent. The U.S. is most notable among the "first to invent" jurisdictions. The majority of patent jurisdictions reward the first to file a patent application and therefore are not concerned about invention date. An interference involves a dispute of inventorship between a patent applicant and another pending application or an issued patent. An interference proceeding is conducted within the Patent Office and relies heavily on laboratory records in establishing the successful claim of inventorship. When the Patent Office issues patents claiming the same subject matter, it is left to the court system to determine which patent is valid.

Evidence provided by someone making a claim of inventorship is considered to be self-serving and therefore looked upon with skepticism. This evidence needs to be corroborated by a noninventor, who has far less stake in the proceedings and therefore is considered to be a more credible source. A well-crafted documentation system is designed to address this need for corroboration.

2.2.3   Defense to a charge of patent infringement

When charged with patent infringement, an effective defense is that the one previously invented had maintained efforts toward public use, publication, or patenting. Laboratory records are useful in establishing invention prior to the patented invention, which is the subject of the infringement charge. Continued research in the technology area of the patented subject matter evidence that the invention was not abandoned, suppressed, or concealed. Evidence of continued research is important since the defense is degraded by abandonment, suppression, or concealment of the invention.

2.2.4   Determination of inventorship

The validity of a patent is jeopardized by improper inventorship. An organization may have an internal dispute regarding inventorship and a patent professional may have to assess whether the contributions of an individual rise to the level of inventorship. A defendant charged with patent infringement may assert a defense that the patent in question is invalid for improper inventorship either through mistake or fraud. Laboratory records in such instances are invaluable in determining the role various individuals played in the inventive process.

2.2.5   Nondisclosure agreement prior knowledge

An organization often agrees to hold the confidential information of others in confidence in order to have an opportunity to evaluate the information in the context of evaluating a future relationship or providing services to the discloser. Confidential disclosures are usually defined by both parties signing a nondisclosure agreement. Confidentiality agreements are synonymously known as nondisclosure agreements and represent a legal contract of rights and obligations for each of the parties. This topic is discussed in greater detail in Chapter 3. If the organization had independently developed the same information that it receives under a nondisclosure agreement and wants to use that information outside the context of the agreement, then the organization must show that it had knowledge of the information prior to receipt under the agreement. The need for such a showing of prior knowledge and/or development typically occurs just after receipt of the information from the disclosing party or after litigation has commenced with respect to violation of the agreement.

2.2.6   Trade secret issues

There are often tricks of the trade that may be essential to getting a process to work properly or a product to have certain properties that confer a competitive advantage. These tricks are usually difficult to discover but trivial to perform once found. An organization usually decides that the obligations of disclosure and the limited term monopoly associated with patents make the best form of protection for these tricks to be retained as trade secrets. Examples of trade secrets include the formula for Coca-Cola®, or starting an enzymatic reaction with a touch from a mustache whisker. Maintenance of a trade secret means that access to the details are on a need-to-know basis, storage is in a protected environment, and there are procedures in place for identifying who gains access to the secret. A competitive organization that hires a competitor employee who has had access to a trade secret may have to show independent development of the secret in the face of a charge of trade secret misappropriation.

A variation on this situation is a former employee who has a noncompetition clause as part of their termination of employment. The date of subsequent work in the noncompetition area may also have to be proven after the fact. Since employment and trade secrets are the subject of state law, the requirements of such a showing vary between the jurisdictions.

A developing area of trade secret disputes is industrial espionage. Both an alleged victim and defendant to such a charge will require laboratory records to support their respective positions. It has become common that a developer of a technology will publish their results or insert into their product a marker that if found in a competitor's offering, is indicative of misappropriation. A nontechnical example is a telephone book having fictitious entries that serve as identifying markers. Conversely, a charge is countered by evidence showing reverse engineering, independent development, or other bases for technical knowledge.

2.3   Types of research records and their usage

The forms available for keeping research records are constantly increasing as computers and personal digital assistants (PDAs) become more popular. Even with computerization of the laboratory, the bound and numbered notebook remains the dominant laboratory record-keeping form. A notebook ledger is the primary source of recordation, with supplemental graphs, images, etc. that are the product of instrumentation being secured thereto. Additionally there are alternate media records that can only be referenced in the principal notebook. These alternate media records include videotapes, audio recordings, film negatives, computer discs, CDs, and DVDs.

Regardless of the form of the primary recordation notebook, there must be a mechanism to insert supplemental data records and reference alternate media records. An exemplary set of procedures is provided for a technologist to record experimental data and a manager to assure that an archival system is in place for proper notebook handling once a book or project is completed. It should be emphasized that alternate procedures should be adopted to fit the situational needs of the individual cases with the understanding that the resulting documents and their handling must be able to convince a trier of fact that for the documents to be falsified would require a conspiracy of unbelievable proportions.

2.3.1   Primary notebook usage

The paper notebook is the place where observations and procedures are recorded in real time. The notebook should also include:

- equipment used, illustrations of the set-ups, delays in setting up or receiving equipment;
- theories for proposed experiments;
- ideas resulting from brainstorming sessions;
- ideas for additional experiments; and
- potential new product uses derived from initial work.

These additions are important since they evidence the date of conception of any invention that result from these prospective remarks. The notebook should represent a single source where all ideas and work relating to a project are found or referenced in the case of alternate format media. Effort should be made to provide objective observations and terms without editorial comment as "good," "bad," "failure," or the like. Features of a notebook that will be given credible weight as legal evidence are provided.

2.3.1.1   Notebook assignment

The management of notebooks is the responsibility of both the technologist and the technical manager. Each person working on a technical project should be assigned his or her own individual notebook. Upon assignment, the notebook is the responsibility of the user until completed and checked in with a record custodian. The notebook is the property of the organization and should be regarded as the most valuable asset with which the technologist is entrusted. Considering the costs of salary, equipment, and consumables used to generate the notebook, the high value attributed to the notebook is justified. Considering the value of the notebook, safeguards for protecting it include at a minimum, securing it when not in use and only taking the notebook off premises under extraordinary circumstances, and even then only after assuring a secure copy exists.

In general, a technologist should have a notebook for each individual project so as to keep different conceptual projects separately. As detailed below, this will facilitate information retrieval at a later date. However, the practicality of handling multiple notebooks suggests that each technologist should be issued no more than two notebooks unless there is a compelling justification.

The transition between a completed notebook and a new notebook relating to the same project should require the surrendering of the old notebook before a new notebook is issued. This procedure precludes a challenge of manipulating the data between old and new notebooks. If a technologist has need for reference to an old notebook, a procedure for doing so is detailed in Section 4.4.2.

2.3.1.2   Entry mechanics

The notebook should have preprinted, numbered pages and be bound. Entries in a notebook should be in nonerasable ink to avoid a challenge or subsequent erasure. For this reason, no pencil notations should be made. While once black ink was considered preferable owing to the inability of xeroxography to make adequate duplicates, this is no longer the case. Information must be legible and wherever possible avoid jargon or abbreviations that can be ambiguous. For the sake of efficiency, a term can be explicitly defined once and an abbreviation assigned that can be used consistently thereafter. The use of loose pages to record data that are then transferred to the notebook is strongly disfavored.

2.3.1.3   Alterations

Notebook deletions should involve a legible strikethrough and should not involve the use of correction fluid or erasures. A page should never be removed from a notebook for the obvious challenge that removal creates not only to the creditability of the notebook itself, but the entire record-handling system. Additions should be dated in another colored ink and/ or with a notation indicating that the comments are additive. However, additions should not disrupt the chronology of the notebook. Additions should reference the page and the number of the experiment being modified.

2.3.1.4   Chronology

The notebook must reflect the timeline of experimental work. The entry is started on a notebook page and the date of entry should be inserted on the top of that page. All entries on a notebook page should be separately dated if entered on days later than that shown at the top of the page.

2.3.1.5   Page completion

Every page of the notebook should be filled. Unused portions of the margins and the bottom portion of the page should include a large "X" or hatch marks. Those portions marked as unused, should not be used subsequently. Each experiment should include a code number to aid another researcher review the notebook without the benefit of the author being present. The coding system should be explained as a preliminary matter at the beginning of the notebook. A coding system may include the notebook number, page number, date coding, or combinations thereof. The technologist should complete the page with a date and signature.

2.3.1.6   Attestation

The first level of attestation is the signature of the technologist author. On a regular basis, such as weekly or biweekly, someone who can read and understand the entries should witness the notebook with the statement "read and understood by [reader's signature]." The witness should not be involved in the particular project or have a professional stake in the fate of the project. Even better, is having two people witness the notebook and not relying on the same person to always witness entries. This standard precludes a laboratory manager to whom the technologist author reports. With a breakthrough discovery, a notebook entry should be witnessed the same day or at least before the next regularly scheduled witnessing.

The reason a witness is remote from the project is that in several types of proceedings the statements of a coinventor are not admissible to prove conception, reduction to practice, or other critical inventive milestones. Avoiding coinventor witnessing saves one from needing to understand the complexity of evidence rules within interference and litigation proceedings.

2.3.1.7   Guest entries

In instances where someone other than the assignee of the notebook makes an entry, there should be an explanation of why the guest is providing an entry. If there is a person who represents a permitted exception to the one-person–one-notebook rule, then it should be noted as a preface in the notebook. The most common instance for a shared notebook is experiments that run continuously over days or longer and shift workers share the data recordation. A guest contributor to a notebook might be someone taking a reading after the notebook assignee has left unexpectedly. A guest entry should be signed and dated by the guest.

2.3.2   Supplemental records

There are any number of spectra, instrument printouts, blue prints, schematics, and the like that one would like to incorporate into a notebook.

These supplemental records should be attached by a method that will not degrade appreciably over time. Many cellophane tapes are themselves unacceptable for this reason. As it is unclear how a particular adhesive might age, the use of glue and tape/staples to give redundant forms of adhesion is preferred. After attachment of a supplemental record, a writing including lines and a date of insertion should be written to overlap the edges. In the event where the supplemental record separates from the notebook substrate, it's location and identity can be determined from the added writings.

When an insert is taken from an instrument output it should be printed with, or have written thereon, the file name, negative number, or other corroborating information. The instrument make and model should also be noted along with the conditions under which the output was generated.

2.3.3 Alternate media references

Some types of data are not amenable to insertion into a notebook in a way that affords evidentiary controls. For instance, a computer readable disk could be inserted into a pocket affixed to the notebook, but not in a way that is not susceptible to a challenge of subsequent tampering. The solution lies in archiving a copy of the alternate media in a way that a custodian not associated with the project in question restricts access to the media. A notebook reference to the media and custodial control completes the incorporation by reference of the alternate media. The requirements of a custodial record-keeper with respect to alternate media are the same as those with respect to completed notebooks, and are described in detail in Section 4.

When the alternate media is computer editable, it is strongly recommended that the custodian be given two copies of the alternate media: a first copy that is available for check out and usage by a technologist and a second copy that is archived. Still better is a system where the duplicate versions of the alternate media are stored in different locations to provide a measure of insurance in case either repository suffers a fire or other catastrophic event.

The question as to what needs to be deposited with a custodian is largely dependent on the nature of the technology, yet the material placed in the hands of the custodian should serve as proof when called upon to show that the invention had been reduced to practice and/or that development was ongoing. Reduction to practice is supported by samples, data, and characterization that evidence that one had possession of an exemplary version of the invention. In the case of a recombinant organism, deposit of the organism itself with a certified repository satisfies this requirement. The storage of test samples evidence proof of new alloy or structure under development. Videotapes of the operation of a production process under various conditions are strong evidence of a production process refinement.

Successful experimental data are not the only information that is sup-
portive of invention. Should one be so fortunate as to get only positive
results, it is still desirable to perform comparative experiments that serve
as controls or benchmarks corresponding to known systems. Comparative
data outside of the inventive scope are also helpful in establishing surprising benefits found within the inventive range. The extent of the effort
needed to discover the operative range of an invention and other such
information becomes quite helpful in responding to a claim rejection or
validity challenge based on the obviousness of the invention.

2.3.4   *Electronic record-keeping variations*

With the advent of computer networks, the notion of the conventional paper
notebook is being challenged. An electronic record-keeping system can now
be put into place that does away with the conventional concerns about ink
color and legible handwriting. These systems, while still in the minority,
offer considerable efficiencies in terms of data-handling and subsequent
retrieval. The key to an effective electronic record-keeping system is that
neither a technologist submitting an entry to such a system nor anyone else
associated with the project has the ability to edit or otherwise modify an
electronic record entry. The computer system manager now has the responsibility for building a set of firewalls and access limitations that will serve as
evidence of a credible data-handling system. The system manager can be
expected to proffer an affidavit or testimony as to the system implemented.
With a noneditable record entry system in place, a system manager provides
an archival backup copy of entries on a removable storage medium that is
amenable to placement in a repository. With a chain of custody from the
technologist onto a server without the ability to edit data after submission,
followed by the backing up of the data records onto portable storage media
that are archived and stored in a controlled access setting, one can achieve a
strong evidentiary trial without a written notebook.

Separate from the storage of data records on portable storage media, a
computer system manager can leave access to prior record postings as
a reference service to the technologist. This system is beneficial in that the
postings can be compiled into a database that is searched by keywords, field
identifiers, cross-references to other postings, and whatever else might be
desirable. Additionally, an electronic record database can also track the
identity of individuals accessing certain data postings and allows for access
restriction to selected personnel.

While there is an expectation that electronic record-keeping will only
expand in the future, it is important to note that patent authorities and the
courts are accustomed to weighing the evidentiary value of handwritten
notebooks. The experience of individual triers of fact with respect to electronic record-keeping and computer systems may be quite limited. As such,
in constructing an electronic record-keeping system one should err on the

side of excess system controls and mechanisms that validate the integrity of the records. While there is no reason that a completely electronic record-keeping system cannot comport with the evidentiary requirements for various proceedings with a patent authority or a court system, such occurrences still remain the exception.

2.4 Record archiving

Adequate record-keeping alone does not preclude a challenge as to the evidentiary value of the notebook. A complete data-handling system prevents technologist access to archived notebooks and thereby removes the possibility that the inventor could alter the recorded data after the fact. The record archive must also serve the need of the technologist or his successor to access deposited data for the repetition of experiments. A cataloging system is in place to retrieve archived items for evidentiary purposes and for technologist referential needs. A series of useful steps are detailed in this section that, when implemented, provide a data chain of custody that will be given credence by a trier of fact.

2.4.1 Technologist–custodian interface

The custodian of records within an organization is someone who could not possibly be an inventor. A custodian is often a technical librarian, office manager, or an office professional. The custodian has a logbook that records an identification number of a notebook checked out to a technologist. The date of the notebook checkout is also recorded since a notebook is in the hands of the assigned technologist until full or a predetermined period of time has elapsed. Typical deadlines for the return of a notebook are 6 months to 1 year.

Before a technologist receives a new notebook, an old notebook relating to the same project must be returned for archiving. When a technologist is requesting a notebook for a new project, the total number of notebooks outstanding to the technologist is reviewed. If the number of notebooks outstanding exceeds a preselected number, the request is denied and the technologist is referred to a supervisor to justify the need for the additional notebook.

The other time a technologist interacts with a document custodian is when there is need for information from a previously archived notebook. The notebook may be used later by the author or another technologist working on a subsequent generation of a project to reproduce experiments, obtain control data, or for numerous other purposes. The custodian cannot release original notebooks without an archival copy being secured. The custodian indexes through a sign-out and sign-in process when the borrowed copy of the cataloged notebook is returned to a technologist laboratory setting.

2.4.2   Audit committee

An audit committee is established to assure that technologists are using proper notebook procedure, technical managers are conveying procedures and enforcing the procedures, and the custodian is retaining isolated archival notebooks and the various custodial logs. An audit committee usually includes the patent attorney to assess the evidentiary robustness of the data-handling system, a technologist, a custodian, and a manager so that all the roles involved in laboratory data recordation are represented. The audit committee should meet on at least an annual basis or when particular issues are raised. The committee has the responsibility for setting and implementing a data-handling system that tracks the current state of evidentiary rules and procedures.

2.4.3   Storage and security

A notebook returned to a custodian is checked for compliance with the record-keeping procedures detailed above with respect to page completion, entry permanency and legibility, witnessing, and supplemental data insertion. Any deficiencies are noted in a custodian bound and numbered notebook. A notebook absent inventor or witness signatures is returned for correction, otherwise the deficiencies are noted and the notebook passed on through the check-in procedure. While it is tempting to correct deficiencies in the notebook as part of custodial check-in, this should not be done. Retrospective changes to the notebook are specifically what this laborious process of safeguarding records is intended to prevent. Invariably having a procedure for regularly modifying notebook entries will be far more damaging to the evidentiary value of the records within the system than the specific deficiencies within a given notebook.

Instead of modifying defective entries, the inventor and the manager are notified of the notebook deficiencies to provide feedback that record-keeping protocol has not been observed. A custodial checklist of possible defects is an efficient way of communicating problems and assures that the remarks remain consistent between multiple custodial personnel and incidents. Repeated problems with the same inventor or common deficiencies become topics for audit committee review. The use of a checklist helps an audit committee track defect statistics and provides evidence of a uniform standard in the event that disciplinarian action is to be taken.

A custodian, upon checking in a notebook, creates a duplicate copy of the notebook. The original or copy is then placed in a secure storage facility having restricted access and an access log.

For large research organizations, it is often most efficient to transfer the notebook or other checked-in material to microfiche to facilitate long-term storage and the noneditable format of microfiche strengthens the evidentiary value of the record-keeping system. The use of a commercial contractor to perform microfiche transfer adds to the evidentiary value by including a separate log of data transfer dates independent of that associated with the research organization.

For a small research organization, a locked file cabinet with access being limited only to the custodian is a viable system. Alternatively, a commercial record storage facility will provide document warehousing in a variety of formats. Archival storage of the original documents with reference access to notebook paper copies preserves the evidentiary integrity of the originals.

A custodian maintaining an original record collection must not only provide security over the records but also the access to the collection. The key, password, or other access mode must itself be secured in a way that supports the evidentiary position that no one generating records has access to original records once checked-in with the custodian. Storage of the access key in a restricted area such as a locked drawer and a procedure to list and limit the personnel who can gain use of the access key, when use is granted, and how the key is retrieved are adequate to verify proper chain of custody with respect to an access key.

2.4.4   Accessing stored information
The reason for storing records is that there is an expectation that the records and the particulars of record entry maybe called upon at a later date. There are two general types of request for stored record access: evidentiary and technical. Stored record access for each of these purposes is discussed.

2.4.4.1   Evidentiary inquiry
The archival data storage system is intended to support evidentiary inquiries made of the research organization in a legal context. Access to the stored information is therefore provided in those instances where proof of an inventive step such as conception, reduction to practice, continued efforts, or a due diligence investigation are required. Typically access to the stored information is made by an attorney preparing for, or involved in, a question about an inventive act or the inventive contribution of an individual. Other organization officials may also need access to establish a variety of facts including employment dates. In some instances, records requested as part of litigation may be subpoenaed and the custodian should be prepared to provide an affidavit or deposition detailing the mechanics of the record-keeping system in place and the chain of custody as to those documents.

The document custodian upon receiving a request for access to the archival materials should log information about the requester and the purpose of the request. An inquiry should be made of the requester to determine if their purpose requires access to the archived material itself, or if access to a verified copy is adequate and the answer noted. The chairperson of the audit committee or another organization official should promptly review the request and transmit a response to the custodian to act accordingly. If at all possible, the original archived document should be retained and a copy provided to once again preclude a challenge that a document was outside of custodial care for a period of time. When the original is required, a certified or notarized copy should be retained in the archive so as to maintain document storage integrity.

When a document or copy of a document from the archive is provided, the document should be labeled with appropriate limitations on the use of the document consistent with the approved request. Labelings include "confidential," "do not copy," or "attorney's eyes only." The latter is typical were a competitor is granted access to documents as part of litigation, but there is a stipulation made prior to forwarding the document that the competitor will not derive a technical advantage through review of the document.

2.4.4.2 Technical inquiry

A technologist may have a need to refer to a notebook that they themselves previously completed or a notebook completed by a colleague. While the evidentiary copy of records should not be accessible to a technologist after submission, an organization has some discretion in deciding the safeguard level applied to information retrieval for technical purposes. Efficiency and trade secret maintenance are the competing factors that are balanced in implementing a technical information access policy. The organizational efficiency extremum is a self-serve library granting a technologist unlimited access to archived record duplicates, while the other extremum involves taking steps to curtail access consistent with protection of trade secrets found within the data-handling system. This is a fundamental policy decision that is intimately tied to the nature of the organization mission and the technology. An organization must decide as an initial matter whether it will use trade secret law as a basis for protecting intellectual property in concert with patents. The factors to be assessed in determining if particular information constitutes a trade secret are difficult to apply prospectively. As a result, a policy to treat all materials as trade secret should be implemented and relaxed as justified by a particular instance, rather than attempting to add restraints after information has received casual treatment.

2.5   Trade secret maintenance

Once an organization makes a determination that trade secrets are a valuable way to protect certain inventive endeavors, then a system of data-handling must be implemented consistent with this objective. It is unlikely that a trade secret can be adequately maintained if it has not been treated as such from inception, since seemingly harmless acts can destroy the secret nature of the information. The details of trade secret law are available to the interested reader.

The definition of a trade secret is relevant to the steps taken for trade secret maintenance. The Uniform Trade Secret Act, §1(4), as a default definition states that "'trade secret' means information, including a formula, pattern, compilation, program device, method, technique, or process, that (1) derives independent economic value, actual or potential, from no being generally known to, and not being readily ascertainable by proper means by, other persons who can obtain economic value from its disclosure or use; and (2) **is the subject of efforts that are reasonable under the circumstances to maintain its secrecy.**" (emphasis added)

This section will detail information-handling procedures consistent with the maintenance of a trade secret. It is important to note that information-handling is only a portion of a trade secret maintenance policy to establish reasonable efforts to insure secrecy. Treatise and the laws of a particular jurisdiction regarding it are instructive in establishing the full set of organizational behaviors consistent with trade secret maintenance. Trade secret policies can be viewed as an added level of care, beyond that needed for retaining an evidence system for patents. A policy to apply for patents and simultaneously maintain trade secrets should not be viewed as being self-contradictory.

Information is more likely to be viewed as a trade secret, if the owner of the information treats the information as if it is valuable and secret. Evidence of steps taken to maintain information as a secret tends to be given weight, consistent with the Uniform Trade Act definition.

2.5.1 Compartmentalized information

The practice of allowing access to information on a "need-to-know" basis serves to compartmentalize information and make it less likely that the departure of any one person having information access to a competitive organization will be able to convey the trade secret with them. In another common setting, suppliers or other outside personnel should only be supplied with redacted versions of documents consistent with their responsibilities.

The document custodian working in a compartmentalized information environment in such instances should be given instructions as information access parameters for various individuals. In most circumstances the delivery of a complete notebook or other research record will contain information beyond that needed for a specific task. The ability to search archived information through notebook tables of content or an index database facilitates delivery of only the needed information to a technologist. In some instances, a technologist will be asked to document a particular process as a series of segmented steps, with each step being bound as a separate recipe that is available only to individuals performing that step. The materials supplied to a technologist should carry a cover labeling the contained information as a trade secret and include in easily readable format the daily handling procedures for the material.

2.5.2   Daily handling

The daily handling of a trade secret is designed to give the impression to the technologist imparted with the information that the organization regards the information as both secret and valuable. One way to maintain such information is to restrict it not only to an essential number of individuals but also to restrict the locations in which the information can be used. Locked facilities, the absence of copying facilities, signs indicating the presence of trade secrets, and the restriction of visitors from those areas where trade secret-containing materials are present, all enhance the information status as secret and valuable.

The handling of trade secret information when not in use by a technologist can be determinative of its ultimate characterization. The storage of such information in a safe or at least a locked desk when not in use is good policy for trade secret record usage. Securing the documents when not in use and a checklist of steps to be followed with trade secret information signed out from the record custodian is satisfactory in showing printed materials containing information that is regarded by an organization as secret. While this text has detailed the handling of trade secret information, there are numerous other aspects of trade secret law that will need to be addressed through consultation with an attorney.

2.6   Concluding remarks

Research records represent the institutional memory of an organization with respect to technology. These records can be drawn upon to provide evidentiary support in a variety of legal contexts relating to inventorship, employment, contracts, and trade secrets. Some typical processes to assure weighty consideration of research records by a trier of fact have been provided. These should not be taken as definitive; instead each contemplated aspect of a record-keeping system should be reviewed against the standard of asking whether this step made it less likely to a trier of fact that records handled in this way were subsequently altered for self-serving purposes.

With this standard as a touchstone and process modifications based on law changes in the relevant jurisdiction, a record-keeping system worthy of the effort will be obtained. Efficiencies realized through electronic record-keeping should ameliorate some of the burdens associated with managing an archival data collection, but at the near term price of uncertainty as to how such systems will be weighed in a legal setting. The use of an audit committee to assess new procedures and compliance with existing policies adds gravity to any system in place. Using a diligent, common sense approach to record-keeping one can expect an implemented system to provide legal and technological benefits that will more than justify the effort.

Additional reading

Record-keeping policies
University of Florida
   http://rgp.ufl.edu/otl/goodrecords.html
Fox Chase Cancer Center
   http://www.fccc.edu/ethics/RecordKeeping.html
Haverford College, Department of Chemistry
   www.haverford.edu/chem/302/general.pdf

Trade secrets
Milgrim, R.M., *Milgrim on Trade Secrets*, 17th rel. Matthew Bender & Co., New York, 1983.
Jager, M.F., *Trade Secrets Law*, 7th rel. Clark Boardman Company Ltd, New York, 1989.

chapter three

Inventor actions that can jeopardize patent rights

Roberta J. Morris
Patent Attorney

Contents

3.1 Introduction

Inventors can jeopardize their own rights to a patent, and even if they obtain a patent, its value may be compromised by actions and omissions they would never suspect could have such consequences. This chapter will explore these serious matters.

The problem arises because an inventor can unwittingly create invalidating prior art in the early stages of inventing or creating a business to exploit an invention. How can this happen? Alas, the answer is: in many different ways.

By this point, you should be familiar with the term *prior art*. You may think of prior art as something that other people create, but you, the inventor, can create prior art yourself, prior art that could invalidate your patent. OK, you say, I understand that if I come out with model 1 of my invention, and then realize I can make some nice improvements, maybe my model 1 is prior art to the patent I want on model 2. You are right, that is one scenario. But another scenario is right there back at model 1. You offer it for sale at a trade show, even though all you have is a prototype of the main assembly, not a working model. If your patent application is filed more than 1 year after the date of the trade show, there is a good chance that your patent, if granted, could be invalid. And that is not the only way you can shoot yourself in the foot.

This chapter, which could be called "Read it and weep," will discuss the various ways inventors can jeopardize their rights to obtain a patent, or their rights in an already issued patent. Another subtitle for this chapter is "Never lie to the patent office." We will discuss the very dire consequences of misleading the Patent Office, and explain why good patent attorneys will want you to be as straight with them as you can. Honesty is always the best policy, and when you are dealing with the Patent Office, it had better be your only policy.

3.2   The concept of the "critical date"

When you apply for a patent, your application receives not just a serial number but also an application date, sometimes called your **filing date**. If you file continuation applications or divisional applications based on this "ancestor" (often called a "parent"), your filing date for those children applications will be the filing date of the ancestor, not the actual filing date of the specific child application. Another kind of child application, a continuation-in-part, may have different filing dates for different (ultimately allowed) claims. This is a complication that is beyond the scope of this book at this point, but it is worth keeping in the back of your mind.

Your filing date could be the date you filed a provisional application. Whether or not our regular application is entitled to the provisional's filing date will depend on whether it supports the claims that are ultimately allowed. This could be a tricky evaluation, or an easy one, but either way you need to do it.

Exactly 1 year before your filing date is your **critical date**. This is a simple arithmetic calculation: just subtract 1 from the year of the filing date. You do not need to worry about Saturdays, Sundays, and holidays when counting backwards. Counting forwards is different. If you, as a potential patent applicant, know of activity that could be critical (in the patent sense!), then, unless your patent application is filed before it, or very soon after it, you will care about your critical date, which will be 1 calendar year before the filing date. If that date is on a weekend or federal holiday, then you must file by the last business day **before** the year expires.

Note: If you know something about patent practice already, you may be thinking "Wait. I know that when the Patent and Trademark Office (PTO) gives me a time limit of, say, 3 months, and the expiration is on a weekend, I have until the next business day to file whatever it is." That is right. But that next-business-day rule applies only to time limits for replying to the Patent Office. It does *not* apply to the critical date calculation. If you know of an event (sale, offer, use, or publication) that could start the clock on the 1-year bar, then make sure your application is filed *before* that date in the next calendar year. For example, if you made the first sale of your new Christmas lights invention on December 25, then your patent application will have to be filed by the last pre-Christmas date that the Patent Office is open for business.

What happens if you did not file before the year expired? It could be bad, and it could be very, very bad. At a minimum, people potentially or actually concerned with your patent, whether as licensees, infringers, or investors, will discover this pre–critical date activity and think your patent's value is very questionable. At worst, you will assert the patent while trying to hide the facts about the pre–critical date activity in court (having already hidden them from the PTO), the accused infringer will be able to prove that your patent is unenforceable for inequitable conduct before the PTO, the accused infringer could win an award entitling it to have *you* pay its attorney fees, and the accused infringer, having won on the patent claim, now has a good antitrust claim against you for trying to monopolize by asserting knowingly an invalid patent, and you could be liable for treble antitrust damages.

Moral: You (and your patent attorney) must always keep in mind the last possible date you can file your patent application so that you do not have to worry about any possible statutory bars that might be lurking in your early sales, marketing, and licensing efforts.

If you have already begun to be patent law–savvy, you may be asking yourself: is this 1-year rule in the statute, the regulations, the patent examiner's manual, or is it judge-made? The answer is Yes! It is in the statute, and there are regulations based on it, and provisions in the patent examiner's manual that pertain to it, and there are plenty of judicial decisions elaborating on it. But here is the statute, so you can see how it is phrased. It is in United States Code, Title 35, Section 102:

A person shall be entitled to a patent unless

***

(b) the invention was patented or described in a printed publication in this or a foreign country or in public use or on sale in this
country, **more than one year prior to the date of the application
for patent in the United States**.... (emphasis added)

Section 102(b), also referred to as the "on-sale bar" or the "statutory bar,"
specifies four triggers that will invalidate the patent (or lead to a rejection of
the patent application) if they occurred more than 1 year before the application date:

1. another patent, U.S. or foreign
2. a "printed publication," U.S. or foreign
3. a "public use" in the U.S.
4. putting the "invention" "on sale" in the U.S.

In this chapter, we will ignore the first item — another patent — because
obtaining a patent is not likely to be the kind of accidental activity that can
suddenly undermine the validity of an inventor new to the patent system.

The other three items — publications, uses, and sales — present the
threshold question: What is the "invention" for "statutory bar" purposes?
Fortunately, that is often an easy question. As long as everyone agrees that
all the claims "read on" (see Chapter 1) the inventor's commercial product,
then there is no reason to pause over the word "invention." However, if
there are differences between the commercial embodiment and one or more
of the various claims in the patent, then the 102(b) issue gets still more
complicated.

Then the thorny questions are:

• what is a printed publication?
• what is a public use?
• what constitutes putting something "on sale" in patent law?

This is the order that the statute suggests. But in terms of complexity and the
likelihood of a seemingly innocent event being held to be a "statutory bar,"
it makes more sense to deal with them in the opposite order: sales first, uses
second, publications third. And before we get to any of that, we have to
consider:

what is "it" that has been placed on sale, used publicly, or
described in a printed publication?

3.3   The "invention" in 102(b)

Placing something "on sale" includes offering it for sale as well as actually
selling it. What is the "it?" "It" is the claimed invention — as you learned in

Chapter 1 — but what shape does it have to be in? If you offer to sell the thing you *think* and *hope* you can make, if only someone will pay you enough, would that trigger the "on sale" bar? The answer is: *maybe*. The standard is that the invention — as it will be claimed when, someday in the distant future, the claim language is allowed by the Patent Office and the patent is issued and you assert the patent against an alleged infringer — must be "ready for patenting."

What does it mean that an invention is "ready for patenting?" Does that have anything to do with an invention being "reduced to practice" — another phrase your patent lawyer may use?

"Ready for patenting" could mean that you have already filed a patent application, or already reduced the invention to practice. But the problem for inventors is that it also can mean less than that. Your invention may be "ready for patenting" (to a court or the Patent Office) even though you see a million things you still have to work out.

"Ready for patenting" means that the conception (the idea) of the invention is completed: you know how and what to do to achieve the purpose you want. In addition, if you have not already built a version of the invention that works, you have at least made some drawings or written out instructions that a person of ordinary skill in the art could follow in order to build one that would work.

"Reduced to practice" means you have a working model. For some unusual inventions (e.g., satellites), you may not need to make a complete prototype that has every element in the claims. But those situations are quite rare. There will be more about "reduction to practice" in connection with establishing your date of invention and "interferences" (discussed in Chapter 4), but for now what you need to know is that if you offer for sale an invention that you have *not* reduced to practice, you may nevertheless create a *statutory bar* on the date of the offer.

If you are about to offer your invention for sale, make a note of the date, and then get your patent application filed before 1 year is out. Also, make sure you tell your patent attorney about the offer, so that together you can evaluate any possible statutory bar events.

3.4 Sales and offers for sale

If something is sold, there will almost always be an invoice with a date on it. That, very likely, provides proof of the date of sale. But the statute also says that "offers for sale" can bar a patent. How do you know when something has been *offered*? The answer is that for once patent law actually uses other aspects of the law. If what you have done would constitute an offer under the Uniform Commercial Code, then it is an "offer" for patent law purposes.

3.5   Sales: the special case of method patents

How do you place a "method" on sale? One way is to sell or offer to sell a
machine that performs the method. In the case of methods that are performed by computer, selling or offering the software can create a 102(b) bar.

3.6   Licensing the patent or the invention

There is some good news in this chapter. Licensing or selling the rights to
your invention does *not* constitute a 102(b) bar. The bar is against selling the
invented thing, not the intellectual property rights. Thus you can offer a
share in your invention to someone who will machine your prototype or
perform services for your start-up company, and neither the offer nor the
consummated sale will constitute a statutory bar.

3.7   Public use

A "public use" can also start the clock for a statutory bar. A typical public
use is showing your invention at a trade show. Even if you have not
commercialized production, and the thing you show is a prototype, you
could still need to file within the year because of the 102(b) bar.

   If the demonstration model has every feature of the claimed invention,
then the claim will "read on" that model, and the model will be invalidating
prior art if it is used in public more than 1 year before the filing date. If the
demonstration model lacks some features of what you will ultimately claim
in your patent, but if those features are already standard for such devices (so
that the person of ordinary skill in the art would have no trouble adding
them, nor any trouble realizing that they needed to be added), then you
could have a statutory bar problem. It will arise because the standard
features of earlier devices, combined with the demonstration model, will
render the claimed invention obvious to an ordinary artisan.

   Who is the "public" in "public use?" The trade show example is easy:
the public is the public. But what if you show your invention to friends who
visit your garage? What if you need to test your invention out in the open or
with real users or under real conditions? The key concept in determining
whether something is "public" is whether confidentiality restrictions are in
place. These can be in writing. For example, you may need some shop work
done on your invention and you may get the mechanic to sign a confidentiality agreement. Or you may be looking for some seed money and you will
ask the potential investor to sign a Nondisclosure Agreement. In some cases,
too, the confidentiality restrictions may simply arise out of the circumstances. Friends being shown your latest invention, especially if they know you
always apply for patents, can be expected to understand that the demonstration is in confidence. If you are thinking about using your invention,

even a prototype or a subassembly, with people who are not in a confidential relationship with you already, you should (1) consider having them sign a Nondisclosure Agreement, and (2) be aware that you may have started the clock running on a 1-year bar.

3.8   Negating public use: experimental use

The good news is that there is a way to "negate" the determination that a use was public. (I choose the word "negate" on purpose: the courts have said that this is not an *exception* but rather a *negation*.)

Inventors are permitted to experiment with their inventions. If the nature of the invention is such that the experiments must be in public, then the use for experimental purposes is not considered a "public use." The classic example, one from the 19th century, was an inventor of a new wooden roadbed. For 6 years before he filed his patent application, he used the roadbed on a tollroad that his company operated. (At the time, the grace period was 2 years, but six is still more than two.) The Supreme Court held that because the invention was a roadbed, and it was necessary to give it a long-term test in a real situation to confirm that it worked for its intended purpose, the use was "experimental" and the patent was not invalid.

You may not be as lucky as that inventor. Among other things, you must be careful that the experiment tests claimed subject matter, not consumer responses to things like marketing, packaging, color, etc. Again, it is wisest to consult a patent attorney about any activity that could start the clock on a statutory bar. Whenever you think that a question could be raised about such activity, the wisest course is to present the information to the Patent Office and let the examiner decide if there is a bar or not.

3.9   Publications

A publication is just about anything in writing. For example, it can be a book, such as a textbook in the field of technology of interest, or a journal article in a refereed journal. Those are the obvious kinds of publications, but a 102(b) publication might also be an advertising flyer, or an instruction manual included with a device when it is sold. It can be something you put on your website, or something you explain in an e-mail if you do not first explain to the recipient that the information is confidential and obtain the person's agreement not to disclose. Brochures and other such material — whether in paper form and available from your company in person or by mail, or electronic and available on your website — can be "printed publications." Even a graduate or undergraduate thesis, provided it is kept in a place and indexed in a manner that makes it accessible to the public, may qualify as a printed publication even though it has never been "published" in the ordinary sense of the word.

Then, too, a "printed publication" does not have to consist of words. A picture — a photograph, a diagram, a flowchart — can be a sufficient disclosure of your invention to the public that it counts as invalidating prior art.

3.10   Reminder: other countries' patent laws are different

The U.S. is unusual in giving people a 1-year grace period to publicize, sell, and publicly use their inventions. Many countries give no grace period whatsoever. If you are in a scientific field where you publish papers regularly, you may have already learned that once you publish you cannot patent in Europe or Japan. If you want patent protection outside the U.S., you will need to learn about these rules.

3.10.1   Strategies for selling and using before applying for a patent

3.10.1.1   Watching the clock
The best strategy for not running afoul of the statutory bar is to know it is there, to keep track of dates, and to file as soon as you can.

3.10.1.2   The provisional application
Another good strategy is to make use of provisional applications. Take the publicity you want to put on your website, or the journal article you have in manuscript and will talk about at a conference, and have your patent attorney or agent file it with the Patent Office as a provisional patent application. When you file your regular patent application, you will claim priority based on the provisional filing, and that will in most cases protect you against your own materials being used against you under 102(b).

3.10.1.3   Confidentiality agreements
Whenever you ask someone's help in building part of your invention, whenever you seek funding, whenever there is any occasion where you could be held to have disclosed your invention to a member of the public, ask that person to sign a confidentiality agreement. By doing so, the person no longer has the status of being "the public" but becomes aligned with the inventor for purposes of 102(b).

3.10.1.3.1   Note: Thieves can invalidate your patent!   The bad news is that if there is bad news, it could be even worse than you imagined. Say that you enter into a confidentiality agreement with someone and then the deal goes sour.

Not only that, but the other person steals your idea, and offers it for sale. That offer can start the statutory bar clock ticking. Even though the offer did not come from you, even though you may not know about it, and even though you may be able to sue the thief for misappropriation and win, you will still need to have filed your patent application within 1 year of the *thief*'s offer for sale. The rationale behind this rule is that once subject matter is in the public domain, it can never be privatized again. No matter how wrongly it came into the hands of the public, the public interest prevents anyone patenting that invention.

3.11 Case studies

3.11.1 The "on sale" bar and co-inventors

Once upon a time, the owner of a medical device distribution house (call him D) and a friend who headed a medical supply manufacturing company (call him M) came up with an improved device. D, the distributing company's owner, knew he could sell these devices to many of his existing customers. M knew his company could manufacture the devices quite economically and efficiently without much retooling. M made a small batch of the devices, and on February 1, 2000 sent them to D. Nobody was thinking much about patents or the on sale bar, and the shipment was accompanied by an invoice. D's company paid the invoice in the ordinary course of business. The batch from M could not be sold directly to D's customers because D's company had to sterilize and package the devices before selling them. On April 10, D made the first sale of these devices to a customer. On March 1 of the following year D and M jointly filed a patent application on the device. Both inventors assigned their patent rights to D's company. Was the patent valid?

The Court said *no*. It did not matter that D and M were both inventors. It did not matter that no third parties bought, or were offered, the devices, until less than 1 year before the filing date. What mattered was that there was an invoice and a standard sale by one of the inventors. The court concluded that the invention therefore was "commercially exploited" by that inventor. The sale ran afoul of the "on sale bar" and the patent was invalid. (Brasseler, U.S.A., I, L.P. v. Stryker Sales Corp., 182 F.3d 888 (Fed. Cir. 1999).) That was not an end of it, though; the patent was later deemed unenforceable for inequitable conduct, and the accused infringer won its attorney fees as well.

3.11.2 "Public use"

A company that had developed a new chain for chain saws asked some ordinary users to test out their chain saws under actual conditions. An initial series of tests was conducted by the engineering department. Records were kept carefully and the chains were recovered by the company as they wore out. The design of the chain was modified as a result of the tests. Later on, another series of tests was conducted by the marketing department. No changes to the design of the chains resulted, and records about the users and their experiences were not maintained for these tests. All the testing occurred more than 1 year before the filing date. Was the patent valid?

The court said *no*, but only because of the second series of tests. The engineering department tests qualified as "experimental use" but the marketing tests were "public use." (Omark Industries, Inc. v. Carlton Co., 652 F. 2d 783 (9th Cir. 1980).)

Appendix

SAMPLE CONFIDENTIALITY AGREEMENT

IN CONSIDERATION of the disclosure which has been, or which may be, made by {Discloser Name}, or any of his agents ("Discloser") to_____or any of its agents ("Disclosee"), of certain confidential and proprietary business information and ideas of Discloser (the "Confidential Information"), and that the Confidential Information is an extremely valuable and important asset of Discloser, and that any unauthorized use of the Confidential Information may cause material economic and business injury to Discloser, Discloser and Disclosee, individually and on behalf of all of its subsidiaries, affiliates, shareholders, directors, officers, employees, contractors, representatives and agents and their respective successors, agree as follows:

1. *Nondisclosure.* Disclosee agrees to hold the Confidential Information in strict confidence and trust for Discloser and not to disclose or otherwise communicate, directly, indirectly or in any manner whatsoever, any of the Confidential Information, the existence of this Agreement, any contemplated agreement or relationship between Discloser and Disclosee, or any contemplated project, development or venture proposed by Discloser ("Project"), except on a confidential and actual need-to-know basis to the directors, officers, financing sources, and/or professional advisors of Discloser for purposes of preparing and/or approving possible future implementing agreements between Discloser and Disclosee. Confidential Information includes oral or written data, reports, proprietary business and financial information, customer lists, products, prices, computer programs, research, inventions, formulae and related information.

2. *Use.* Disclosee agrees to use the Confidential Information for the exclusive benefit of Discloser and/or evaluating future agreements with Discloser, and for no other purpose whatsoever. The Confidential Information, regardless of its form, is, and shall always remain, the sole and exclusive property of Discloser. Reproduction of the Confidential Information is not allowed unless agreed to in writing by the Discloser.

3. **Return.** Disclosee agrees to return to Discloser, immediately upon demand from Discloser, and not to retain, all Confidential Information disclosed to Disclosee, including, but not limited to, all originals, copies, reproductions, summaries, analyses, interpretations, notes, and/or related items made of or from the Confidential Information.

4. **Term.** The obligations of the Disclosee under the terms of this Agreement shall expire five (5) years from the date hereof or three (3) years after the last disclosure under this Agreement, whichever is later.

5. **Circumvention.** Disclosee agrees not to circumvent or otherwise try to avoid, directly, indirectly, or in any manner whatsoever, any of the terms or provisions and/or the intent or purpose of this Agreement. The terms and provisions of this Agreement are to be interpreted as broadly and liberally in favor of Discloser, and against Disclosee, as legally permitted.

6. **Exceptions.** Disclosee need not maintain the confidentiality of any information:
   a) Which is rightfully available to the public prior to the time of disclosure to Disclosee by Discloser;
   b) Which is rightfully in the possession of or known by Disclosee prior to disclosures to Disclosee by Discloser;
   c) Which was developed independently by Disclosee without reference to the Confidential Information and prior to disclosures thereof to Disclosee by Discloser;
   d) Which Discloser specifically consents to in writing and in advance of any such disclosure by Disclosee; and/or
   e) As required by court order as long as Discloser is notified in writing fourteen (14) days in advance of such disclosure.

7. **Termination.** Either Discloser or Disclosee may, upon thirty (30) days written notice to the other party, terminate this Agreement with respect to disclosures made after the effective date of such termination, provided, however, that such termination shall not affect the Disclosee's obligations relative to Confidential Information transferred prior to the effective date of such termination.

8. **Jurisdiction.** This Agreement shall be governed by and interpreted under the laws of the State of _____ and may not be superseded, amended or modified except by written agreement between the parties. Any legal action regarding this Agreement shall take place in the State of Michigan by a court of competent jurisdiction.

9. **Remedies.** Disclosee agrees to pay Discloser for all damages and expenses incurred by Discloser as a result of any actual or threatened breach hereof or interpreting the terms hereof, including, but not limited to, compensatory, incidental, consequential, and exemplary damages, court costs, expert fees, and attorneys fees. In the event of any actual or threatened breach of this Confidentiality Agreement, Discloser, as the sole beneficiary hereof, may enjoin

any such actual or threatened breach, by ex-parte proceedings or otherwise, if circumstances so warrant, at the sole discretion of Discloser.

10. *Integration.* This Agreement constitutes the entire understanding between the parties and supersedes all previous understandings, agreements, communications and representations, whether written or oral, concerning the treatment of information to which this Agreement relates.

ACKNOWLEDGED, AGREED AND ACCEPTED:

{Typed Name of Discloser}

By:_____

Dated:_____

"DISCLOSER"

{Typed Name of Disclosee}

By:_____

Its:_____

Dated:_____

"DISCLOSEE"

chapter four

The invention disclosure document: recording the essential facts of your invention

Judith M. Riley
Gifford, Krass, Groh, Sprinkle, Anderson & Citkowski, P.C.

Contents

4.1   Introduction — the necessity for complete and accurate information

As a scientifically trained person, you have learned the necessity of keeping good records. Who can forget those days in our first college lab classes when we had ingrained into us the habit of keeping a lab notebook to record every step of every experiment we performed? Any student who approached this task too lightly soon discovered the results of sloppy record-keeping habits; without meticulous and organized recordation, all the care we took to weigh, to measure, to set up our equipment, to follow the steps of the procedure precisely and accurately could come to nothing because we did not have a complete factual record of what we had done. How could anyone reach any sensible conclusion from an incomplete and chaotic mess of scribbling?

At this point in your career, you have most likely been keeping good lab notebooks for some time. You will find the information contained in them an invaluable source when the time comes for you to prepare a document that patent-savvy professionals commonly refer to as an "invention disclosure" document. Bear in mind that however complete your lab notebooks may be, they will not contain *all* the information about your inventions that you will need to convey to your colleagues or superiors, and eventually to the patent professionals who may well turn what you have invented into a full-blown patent application.

Nor does the necessity for good record-keeping and complete information about your inventive activities end with the filing of a patent application or even with the grant of a patent. You may be called upon to provide additional information at any stage during the prosecution of the application or subsequently during the life of the patent. Just to take a few examples: a dispute may arise with someone else claiming to be the first inventor of what you have invented. This dispute may trigger a Patent Office procedure called an "interference," and critical to its resolution will be the key issues of when (we are talking here about the exact calendar date) the respective inventors first conceived of the invention in question, when they first reduced it to practice (i.e., made it actually work), how diligently they worked on it in the meanwhile, and so forth. Or, after your patent has been granted, someone who wishes to practice your invention without your consent may challenge its validity by asserting that your invention was on sale or otherwise publicly known more than 1 year before you applied for the patent, which, if true, would invalidate your patent.

Another possible outcome for your invention could be a decision not to apply for a patent, but to keep what you have invented as a trade secret. In other words, instead of filing a patent application that discloses to the rest of the world exactly how to practice your invention, you or your employer may decide that a better way to maintain a monopoly over the invention would simply be to keep it secret. Even in this case, the confidential records that you keep may be instrumental should a dispute arise with a competitor who applies for a patent on the same invention, or with someone who has managed to illegally acquire the trade secret and is using it to compete with you.

The above scenarios merely scratch the surface of the range of issues that can arise around inventions, patents, and patent applications, all of them heavily dependent on being able to show in detail each step of your inventive and commercial activities and exactly when each important event occurred. By far the easiest and best way to show this is by documentary evidence — lab notebooks, drawings and sketches, notes and letters to others and, best of all, the content of a well-drafted and complete invention disclosure document.

4.2   What an invention disclosure document is . . . and is not

Because corporations, universities, and other institutions have become so much more aware in recent years of the power of patents to generate income and control the flow of commerce, more and more of them have instituted standard procedures for uncovering significant inventions of their employees and delivering information about them to a patent professional in a timely and orderly manner. Hence, the increasing presence of Invention Review Committees, Patent Liaisons, in-house patent seminars, etc. By far the most powerful and popular of these techniques is the Invention Disclosure document.

Broadly speaking, an adequate Invention Disclosure document has three primary purposes:

1. It provides complete enough information so that your organization can make an intelligent decision on whether the invention disclosed therein is significant enough to justify the time and expense it will take to present it to a patent professional for search and subsequent application or perhaps to maintain it as a trade secret.
2. It enables the patent professional first to decide whether the invention disclosed therein is a good candidate for patent protection and then to prepare a complete patent application.
3. It contains a complete enough record of the facts surrounding the making of the invention to resolve the kinds of disputes (as well as others) mentioned in the Introduction.

As a patent attorney, I have seen many standard Invention Disclosure forms cross my desk. Some of them have been good documents that managed to cover all the bases without going into undue detail or becoming repetitious.

Many others have failed in one or more important respects, and this has
been due in some cases to a failure in the standard form itself (e.g., the
invention is described quite well, but the form does not have a place for
such significant information as the date of conception) or very often because
the inventor(s), like our naïve freshman lab student and her notebook, did
not think it important enough to take the time to fill it out completely.

When do you need to fill out an Invention Disclosure form? Depending
on your rank in your organization, this is a decision you might have to make
yourself, or perhaps someone else will instruct you to do it. Of course, no
one will ever ask you to author one if you do not communicate what you are
doing to others. Many inventions are lost every day because the people who
make them do not think they are important enough to bother their superiors. On the other hand, a continual stream of "bright ideas" that leads
nowhere can get you a reputation that you would not care to have. So some
judgment is required.

In general, your organization will be most interested in inventions that
touch the heart of whatever task you have been assigned to do. You are also
likely to see great interest taken in inventions that may be somewhat out of
your area of particular specialty, but bear on other important work in which
the organization is engaged. More rarely will an employer be very interested in something completely outside its field. Obviously, inventions that
represent a significant advance over the way things have been done in the
past are particularly good candidates for further inquiry. Best of all are
inventions that are likely to lead directly to commercially successful enter-
prises. On the other hand, even a clever invention that is too specialized to
have broad applicability to the marketplace is much less likely to generate a
lot of enthusiasm.

To some degree, how far you should go in disclosing an invention will
depend on your organization's general stance toward the importance of
protecting intellectual property rights. If your employer has no history of
ever patenting anything, then you are going to face an uphill battle in
getting someone to listen. But if there is an in-house patent staff, or a person
assigned to interface with outside patent professionals, or a patent review
committee, or the like, then your organization probably already has established standard procedures for getting employees' inventions to the attention of the decision-makers and that will make your task much easier.

So let us assume your employer possesses "patent smarts" and already
has in place a standard procedure for disclosing inventions. At the first
stage of disclosure — simply telling a coworker or superior of your invention — you should make a broad cut. It is better to disclose even things that
seem too trivial to be real inventions than to overlook what may turn out
later to have a significant bearing on future operations. When it comes to
undertaking the not-so-small task of preparing an Invention Disclosure
document, however, this is something you should reserve until you have
got some positive feedback on your invention. A lot will depend on the

particular procedures your organization has in place, so make sure you familiarize yourself with them before you get too deeply into the process. Some organizations encourage early, frequent disclosure. They generally use a "short form" invention disclosure, usually only a page or so long, which may be filled in quickly. Often such preliminary disclosure will be followed by a more detailed disclosure prepared after the short form has been reviewed and the invention therein deemed to be of particular interest.

When you do finally get to the point where you have your organization's standard Invention Disclosure form in front of you and you are faced with the task of turning it into a useful document, it is wise to keep in mind a few things that you *should not* do. You are not writing a research paper for publication in a professional journal. Those reading your Invention Disclosure Statement (IDS) are very likely to be less knowledgeable about your particular specialty than you are, so be careful not to write "over their heads." Of course, the disclosure is not a lab notebook that will be read primarily by yourself, so take care to communicate in enough detail and also in Standard English so that those who read it will have a full understanding of what you are doing. Resist the temptation to be cryptic because it will probably result in a request for further explanation. Doing it right the first time will save you and everyone else time and trouble.

Finally, try to avoid simply repeating yourself as you fill out the various sections of the form. In a well-drafted standard Invention Disclosure form (and unfortunately not all fall in this category), each section is there for a particular purpose and requires you to give additional information. You will usually be asked to explain the conditions you faced that motivated you to make the invention (this section may be called "background," "prior art," "what led you to make the invention," etc.). You will need to make a concise explanation (a "summary" or "abstract") of the invention, as well as a more detailed description of how to practice it and why and how it works. You will want to include sketches and drawings that help explain the construction and function of your invention, and please try to make them legible enough to be understood. You will often be asked to identify any previous disclosures that might have a bearing on the present one — an important recognition of the reality that you most likely work on ongoing projects — and each incremental improvement adds up to a coherent body of your work. You will also need to list any background material (texts, journal articles, issued patents, etc.) that you feel have a bearing on your invention or help others understand it.

You may also expect to find requests for very specific data, such as the date you first thought of the invention, as well as the date when you finally completed it, both conceptually and in practice. You will need to identify all of the coinventors, if any, of this particular invention and what contribution each one made to the whole. In cases where there are coinventors (a good lot of the time), collaborating with them on preparing the Invention Disclosure document is a virtual necessity. You also may be asked to provide examples and test results that show the advantages of your invention. Here is where good lab notebooks prove invaluable.

Some disclosure forms will ask you to list the names of anyone whom you have already told about the invention. You also may be asked to list any steps that have been taken toward commercialization, and most particularly of any actual offers for sale (along with their dates) that may have already been made. You will have to sign and date the Statement, and your signature should be in front of at least one witness. There will usually be a signature and date line for the witness, as well as for yours. The entire document should be marked "CONFIDENTIAL." If this is not on the printed form, then you should add it.

One final cautionary advice in filling out the form — do it promptly, even if you think you are too busy. Intellectual property rights are very time-sensitive and even what you may think is a short delay can result in their loss. Do not let the form sit around on your desk for weeks on end until you find the time. Make time in your schedule right now and just do it. Above all, do not let the process intimidate you. As between a highly polished disclosure document that has taken its inventor 2 months and 17 rewrites to produce and a bare-bones, on-the-fly production that manages to contain everything important and winds up where it belongs *promptly*, there is no question which is more valuable both to you and to your organization.

4.3   Preparing an invention disclosure document

4.3.1   Preparing a standard form

The forms that different organizations use for invention disclosure are as various and different as the organizations themselves. Some are more tightly formatted and contain quite a few "fill in the blank" sections. Others are more free and permit more creativity and individual input. Some request quite a bit of ancillary factual information, while others focus almost exclusively on getting a complete description of the invention. There are also those formatted more or less like a formal patent application and even including a section labeled "claims." We will take a look at some typical formats given in the Appendix. The particular format your organization uses may or may not contain a place for all of the types of information we will discuss below. However, it would be a mistake to skip over a heading or two because you do not happen to see one like that on the particular form you are using. You should at least be aware of the possible types of information some organizations routinely request from their inventors. Possibly, you might decide to add a few items to your document even if the form you are using does not mention them. For example, some disclosure forms unfortunately do not include a place to describe any sales, offers to sell, or other commercial activities involving your invention. That is fine, of course, if there have not been any, and that ideally would always be the case. But sometimes these activities will have already occurred. If you know of any, you should definitely describe them in the disclosure document you are preparing.

It is also possible that your organization does not have a standard disclosure form, or that you are an independent inventor without access to a standard form. You will still need a complete disclosure of your invention in writing if you have any intention of investigating the possibility of securing patent protection over your invention. By reviewing the subsequent subsections and examining the actual examples I provide later on, you should have no trouble putting together your own tailor-made invention disclosure, and doing so will prove an invaluable investment of your time because it will be of immense help to you in your eventual dealings with the patent professionals who will help turn your invention into a patent.

4.3.2 Background of the invention

This section will sometimes be called "Prior Art," although that term identifies only one of the items that properly belong in a good background statement. You will need to explain exactly what prompted you to make this invention — the "problem" you faced, as well as approaches taken by others in the past to solve this problem. Of course, these previous attempts have not completely solved the problem (if they had, you scarcely would have been motivated to make your invention, would you?), and that is why your invention is needed — it does what the prior art has failed to do. Or it does what the prior art does, but does it better or faster or more efficiently or more inexpensively or whatever other advantage you believe your invention has over the prior art.

4.3.2.1 What is the problem?

Considering that you are the inventor, you would think this question would be very easy for you to answer. "We needed a more durable coating for precision tool parts because customers were complaining that the coatings we have been using flake off in certain high-temperature environments." "Existing surgical instruments for performing retinal procedures often cause damage to the vitreous humor and subsequent degradation of the visual field." "While there are a number of folding pocket chess sets available in the market, all of them are too small to handle easily and impossible to stabilize when someone wants to play chess in a moving vehicle." "Automotive paints and coatings are usually applied in a booth or other type of confined area, and this makes it especially important that the amount of toxic vapors they emit be greatly reduced."

All of the above are perfectly straightforward and provide a good take-off point for describing in greater detail what is wrong with the prior art —

the coatings were formulated for lower temperatures; the surgical instruments were designed without realizing what damage they might inadvertently do; the chess sets were successfully miniaturized, but ignored the problems miniaturization would entail; the automotive paints were formulated in an era when the adverse effects of toxic emissions were not as well understood.

Fortunately, most of the inventions you will disclose in your career will solve one or more problems broadly similar to the above examples and you will find it easy to write a statement of what problem(s) your invention solves. But sometimes things are not quite so simple. Some inventions are so pioneering in nature that they do far more than solve a problem. Rather, they give birth to a whole new area of technology. Inventions such as the movie projectors, the laser, the vulcanization of rubber, and so on fall into this category. Often enough, inventions are made whose full implication even the inventors do not grasp. Thomas Edison, for example, thought of moving imagers as suitable only for individual viewing and simply ignored anything to do with projecting them to a mass audience until others had already established themselves as formidable players in that field of technology. Then there are those inventions that are made more for sheer novelty, personal amusement, or user convenience. Does a drinking straw that plays a tune as you suck on it solve any kind of a problem? Yes, in a sense, because it addresses the public's (and particularly children's) constant need for diversion and amusement. The mere fact that there isn't a technical problem in the strictest sense to be solved makes it no less of an invention.

If you find yourself in the situation where you cannot readily put into words what prompted you to make the invention, try putting yourself into the same frame of mind as when you first started thinking about it. Where were you? What were you doing? What were you thinking about at the time you made the invention? Did something "pop" into your mind, or did it require some analysis? What was the *first* thing you thought about after your invention became clear in your mind? What prior approaches had you made and rejected in favor of the invention you actually made and what prompted you to reject them? From all of this may emerge some clues that will help you focus on exactly *why* you made this invention, and that is really the crux of what you are attempting to explain when the question is asked "What problem does the invention solve?"

4.3.2.2   How have others solved it in the past?

This is where you are going to describe the closest prior art of which you are aware. Depending on the sophistication and complexity of the technical field of the invention, this description may be brief and informal ("many nonrefillable toner dispensers for xerographic machines are in use — one example is manufactured and sold by the ABC Corporation under the name 'Tone' "). Or it may be quite lengthy and involve the citation of journal articles, standard handbooks and treatises, U.S. patents, foreign patents, product literature, and the like. Sometimes it may be as simple as an informal account of what workers in that field have always done to solve the problem — maybe they improvise a tool belt to hold specialized tools, for example, and you have invented a more convenient one that can be mass manufactured.

Whatever the prior art is and however you proceed to describe it, make sure that you include everything you know that could have a bearing on your invention. Omitting something because you are afraid it might block you from getting a patent is particularly a stupid idea. Chances are that whoever does a patentability search on your invention will find it, and even if he or she does not, the Patent and Trademark Office (PTO) examiner in charge of your patent application probably will, and all the time and money already spent on securing patent protection may well go right down the drain. You certainly do not want to find yourself held responsible for an outcome like that.

On the other hand, describing or listing every patent, journal article, and commercial literature you have ever heard of that touches, however remotely, the technical field of your invention is an equally ineffective approach to the task. Disclosing too much is virtually the same as disclosing too little because you have created too big a forest for anyone to locate a particularly important tree. Again, you must exercise some judgment to avoid the twin traps of disclosing too much or too little.

The other important reason you need to disclose prior art as completely as possible is that it will be of immense assistance to the patent professional who searches your invention and eventually drafts and prosecutes your patent application. In fact, the same could be said of most of what you put into your disclosure document — the more complete a job you do, the better job you can expect from your patent attorney.

4.3.3 Description of the invention

This will be the heart of your disclosure document. In it you will provide all the details you know about how your invention is made, what components and materials go into it, how it works, how it is used, for what purposes it is used, and who is likely to use it. For most inventions, you will want to include some accompanying sketches or drawings to show exactly what you mean. These need not be elaborate or of professional quality. Engineering drawings and blueprints can be extremely helpful in showing the detailed features of the invention. You may also have graphs or charts that show how well your invention performs in practice. Sometimes, if your invention is a machine or part of a machine, these may be generated while your invention is working. In other cases, you will create them yourself from data you or others gather and record. You may want to include test data that show how well your invention works compared to the prior art. Photographs and photomicrographs may be available and helpful. If you have run experiments in the course of developing your invention, then you will want to include them. Similarly, and this is particularly true of chemical composition inventions, you may want to describe several examples of implementing your invention, and you will need the exact chemical formulae and nomenclature, the process parameters, and any analytical data you have obtained.

Much of this kind of material can simply be photocopied and attached to your disclosure document. Make sure that you reference such attachments with clear letter or number designators.

4.3.3.1   Why is your invention a better solution?

We have already discussed what should go into the prior art portion of your disclosure. In it you identified what problem your invention solves, as well as what others have done before you to address the problem. A natural transition into the description section of your disclosure is a brief explanation of why your invention solves the problem more effectively than anything in the prior art. "I have found through experiment that adding trace amounts of silicon nitride gas to the mixture of gases used in the vapor deposition process to coat our tools greatly improves the stability and performance of the coatings at high temperatures." "Adding magnetic weights to the bottoms of miniature chess pieces makes them much easier to handle and also greatly improves their stability when they are used in conjunction with a metal or metalized playing board."

You will notice that both of the above touch upon the deficiencies in the prior art discussed in the background section of the disclosure document and also tell us in very brief fashion the inventive concept that will be set out in greater detail in the ensuing section of the document. This makes them ideal transitional phrases that help the reader move smoothly from your background statement to your description of your invention and that make your disclosure much more readable.

4.3.3.2   How is the invention made?

An invention, of course, does not have to be an article or thing, although a large number of patented inventions fall in this category. We often see patents covering article of manufacture (a windshield wiper, a pill dispenser, a toy airplane) or chemical compositions (a chemotherapeutic agent, a paint formulation, a pesticide). But an invention can also be a method of doing something (forming laminated flooring material, performing laser surgery on the retina, booking an airline flight on the Internet) or a method of using something (laying the laminated flooring material, administering the chemotherapeutic agent, operating the windshield wiper). Your invention will certainly fall into one of these categories, but it also may fall into more than one category. Take the example of the laminated flooring. Does the invention lie in the floor covering, itself, or rather in the way it is made?

In this case, we do not have to choose one category to the exclusion of the other because it is fairly obvious that the invention lies in both what it is and how it is made. Keep this in mind when you are drafting your disclosure. If the decision is ultimately made to apply for patent protection over your invention, the patent professional will draft claims over every aspect of the invention that he considers patentable, and this may well include not only what the invention is but also how it is made *and* how it is used. You should try to cover all of these bases in drafting your disclosure if you think they are at all applicable.

If your invention is an article of manufacture or a chemical composition, the question "How is the invention made?" is fairly straightforward and not too difficult to answer. You will describe all the steps that are performed in making the article or composition as follows:

1. You will first need to describe the starting materials. A complete disclosure should include both the general categories of the materials you start with (a heat-deformable plastic, an electrically conductive solder material, a length of hollow metal pipe, a reducing agent), as well as the particular materials you have actually used (¼ inch stock polyethylene sheet, epoxy cement, ½ inch diameter copper pipe, a 20% hydrogen peroxide solution). You can even include the brand names and manufacturers of these materials if you wish, but you should also include the chemical name or some other generic description of each brand name material since the Patent Office will require this information.

2. Next, you will describe how you prepare and process these starting materials (cleaning the surface of the stock with a dilute acid, heating the solder until it is flowable, cutting the copper pipe in 8-inch sections, diluting the hydrogen peroxide) to form the various parts and constituents of your invention.

3. Finally, you will describe how and in what order the parts are assembled or the various chemical constituents are combined and processed. Frequently, there will be intermediate steps to describe because many articles of manufacture are assemblies made up of several subassemblies, and many chemical compositions are compounded from intermediate compositions. If your invention is a chemical composition or class of compositions, you will also need to describe in detail the process parameters and conditions (temperature, pressure, time, etc.) under which you formulated the composition, because often these fine details will determine whether you eventually are able to obtain patent protection over your invention.

We discussed lab notebooks earlier and here is where a good set will prove invaluable to you. If you have in the pages of your lab notebook all of the details discussed above, then your task will be that much easier. You will need only to turn them into a coherent narrative. Do not forget to include or attach any sketches, engineering drawings, schematics, printouts, photographs, test results, charts, etc. that you think are important.

4.3.3.3   How does the invention work?

As mentioned above, method or process inventions fall into two broad categories: (1) methods of making something (either an article or a composition); or (2) methods of using something (again, either an article or a composition). If you have followed the steps outlined above, you will have already described not only what the invention is but also how it is made. Still, you have not really finished your description because you have only got to the point where you have the invention lying on a table in front of you. What is anyone supposed to do with it? This may seem like an obvious question to you because you already know what the invention is for and what should be done with it, but do not forget that whoever reads your disclosure might not know, and this is particularly true of the patent attorney or agent who may have the task of turning your disclosure into a patent application. Always assume that your reader does not know anything about the technical field you work in and provide him with a brief, but complete, description of how the invention is used. Again, you will do this step by step, describing the environment in which the invention will be used, the various components with which it will interact, what it will do to these components, how it will affect its environment, and what will be the end result — what will the invention ultimately accomplish and how will this benefit the user?

These general comments will work for most disclosures, but there are categories of method or process inventions that require special treatment. For example, your invention may be a computerized system and employ specialized software that programs the computer to execute a sequence of steps to perform a particular function — e.g., processing insurance applications. Or perhaps the computerized sequence is just a part of a larger manufacturing process, such as quality control, and is used to calibrate and guide other precision machinery or measuring devices. In either case, you should try to include in your disclosure at least a rough flowchart at the systems level showing the sequence of the programmed steps.

Particularly in high-tech fields such as optics, instrumentation, cryptography, etc. you may employ one or more mathematical algorithms in performing the method of your invention. You will need to disclose these equations in meticulous detail, and, if novel, you will probably need to describe how they are derived.

4.3.3.4   What is the kernel of the invention?

The kernel of the invention — sometimes also described as the "gist" or the "heart" of the invention — is nothing more than a succinct statement of what you consider to be the most important thing about your invention. In many cases, it will be the answer to the question "What does this invention do that has not been done before?" Or, if you are working with an already well-developed technology, it might be the answer to a slightly different question: "What does this invention do better than the prior art does?"

Let us try a familiar example to see if we can pull out the kernel of the invention. We have an invention comprised of a planar, horizontally disposed member of a size and strength capable of supporting the nether part of a human torso when that torso is resting on its upper surface. From its lower surface project three elongated members of approximately equal length that are equally spaced from each other in a pattern that approximates an equilateral triangle. These elongated members support the planar member so that it is maintained in its horizontal orientation when the invention is placed on a surface such as a floor, and also space the planar member in a spaced relationship from the floor so that someone can comfortably sit on it.

From this description, you have probably guessed that the invention is a three-legged stool, but we have not quite captured the gist of the invention. Why make a three-legged stool rather than one with four legs? Well, anyone who has sat on four-legged stools and chairs knows that many of them rock back and forth because the legs are not of equal length. But a stool that has three legs will never do this, even if the legs are not equally long because, as we know, three points define a plane. The upper ends of the legs will define the plane of the seat member and their other ends will always firmly rest on the plane of the floor.

If asked to explain the kernel of this "invention," we could say something like "This invention is a three-legged stool and provides its user with a secure seating surface that never rocks or shifts." Of course, your invention might not be so easily captured in a simple statement like this. You may have to use several sentences, and in doing so, you will often find yourself very briefly describing the invention, explaining a bit of how it works or what it does, and maybe even revealing why it is better or different than everything else out there. You will notice that our statement of the kernel of the stool invention actually does all three things.

Do not spend too much time agonizing over this issue. You no doubt have had in your own mind a firm idea of what you think your invention is. You will probably find your initial thoughts becoming more refined as you work through the various sections of the disclosure document. All you need to do is put this down on paper and that will be the kernel of your invention.

If the standard form with which you are working asks you to state the kernel, gist, or heart of your invention, you will usually be asked for this statement either at the beginning or the end of the description section.

4.4   Who is likely to use the invention?

Occasionally, you will find this question in a standard disclosure form, probably for the purpose of deciding whether the invention is significant enough commercially to merit subsequent patent search and application. Obviously, the more categories of potential users you can uncover for your invention, the more marketable it will be. This kind of question gives you a chance to tell your superiors or employer about the importance of your invention. Even if you are an independent inventor, it is still a useful exercise to answer it because it will help you with the decision you must make on whether to invest the time and considerable outlay of money that it will take to try to turn your invention into a U.S. patent. You may have the cleverest invention in the world, but if the potential users number in the dozens rather than the thousands (let alone millions), then there is much less likelihood it will ever pay for itself in terms of patenting, prototyping, and otherwise getting it ready for the market.

Try to think in very broad terms when you frame your answer. As we all know, technology is a moving target. In some fields, the target moves so rapidly that it is hard to even guess what might be the state of the art several years down the road when your invention will be likely first to hit the marketplace. In fact, one problem with a truly pioneering patent is that it is so far ahead of the general state of technology, it may be ready to expire before it can be commercialized. Chester Carlson's first patent on the xerographic machine expired years before photocopiers were close to the point of being mass produced. The same is true of some of the pioneering patents in television broadcasting and receiving.

Hopefully, that particular scenario will not be the case with your invention. At a minimum, you should be able to identify the broad class[es] of your potential users — be they ordinary consumers, the military, research institutes and universities, intermediate processors and fabricators, hobbyists, etc. Of course, you may have an invention that spans both the retail and wholesale market, or one that appeals to several broad classes of consumers. If your invention is an inexpensive or throwaway consumer item — the drinking straw that plays a tune — then it is going to require a very broad class of potential users to make it commercially viable. But if it is more in the nature of, say, a sophisticated analytic machine, then even a handful of potential users may be enough to support a profitable enterprise.

4.5   Publications, sales, offers to sell, and other public disclosures

A standard disclosure form may also include a section, which asks you to list all activities that have been taken to commercialize the invention, as well as any publications that describe or disclose the invention, and any public use of the invention. However, many standard forms do not ask for this information. The reason may be simple inadvertence, or perhaps the result of a deliberate decision not to go into these areas on the disclosure form.

Under the system of patent law we have in the U.S., a number of activities surrounding an invention may forever bar the inventor[s] from obtaining patent protection. These fall into three broad categories: (1) actual sales and offers to sell; (2) a printed publication which describes the invention; and (3) public use of the invention. However, unlike some other countries, the U.S. does give the inventor a full year from the date when any of these activities occur to file a patent application without any loss of patent rights. Thus, if you take the prototype of the invention you have been secretly working on in your shop to the Gigantic Corporation on July 21, 2004 to see if there is any interest in buying it or, perhaps, manufacturing and selling it, then you have until July 21, 2005 to get your patent application on file. The same is true if you write a journal article describing your invention and it is published on July 21, 2004. You must have your application on file on or before July 21, 2005 or you will forfeit your right to a patent. A note of caution: many journals now electronically publish articles months prior to printed publication. Electronic publication dates should be tracked carefully as this date of publication, rather than the later printed paper publication counts. Electronic publishing is a technological innovation that most likely will be considered as falling within the language of "printed publication." Publicly demonstrating your invention at a trade fair on July 21, 2004 will also trigger the same July 21, 2005 deadline for filing your patent application.

There is nothing more frustrating for a patent attorney to receive a complete and nicely written invention disclosure document, do a patentability search, render an opinion that the invention is likely to be eligible for patent protection, draft and file the patent application, prosecute the application through the USPTO, have the client receive the patent, and all the time be totally unaware that the client was trying to sell the invention for several years before the patent application was filed. The nasty surprise of discovering this information often happens when a competitor has decided to sell a product that infringes the patent, and the patent attorney has sent the competitor a letter asking her to stop. Even worse are the cases where the damaging sales activity comes to light right in the middle of a patent infringement lawsuit. What vast amounts of time and money have been wasted pursuing and trying to enforce patents that were invalid the day they were issued because the client either did not know enough to disclose prior sales, publications, and public uses, or, even worse, deliberately concealed this activity from the patent attorney retained to represent his interests.

We patent attorneys can only do our job properly if we are told all the facts, and this includes everything that may bear on the patentability of our clients' inventions. You need to tell us every step taken to commercialize and publicize the invention, and we also need to know of any publication that even mentions it. That is the only way that we can render you an informed opinion on whether your invention merits the time and expense of applying for a patent. It may well be that the events that have already occurred will not actually bar you from obtaining a patent — e.g., you may have installed a prototype of your invention in a customer's facility for the purpose of determining how well it works in a real-world situation. Activity of this type may possibly be considered an "experimental" use and not true commercial activity or public use. However, and as you can well imagine, this particular area of law is extremely complex, and it takes our specialized knowledge to render you an informed opinion. Obviously, we cannot do that if you decide, for whatever reason, to keep us in the dark.

In the event you are aware of activities that might bar you from patent protection over your invention and you find no place for it on your disclosure form, you are not relieved of your obligation to disclose it. Should your invention generate enough interest that your disclosure is given to a patent professional for a patentability search and further proceedings, then you *must* disclose to him or her all the information you have as soon as possible and let him or her make the decision on what to do about it.

4.6   A few examples

Example 1 through Example 3 illustrate some typical institutional invention disclosure forms. As you can see, they vary in their complexity, and also in their level of formality. Although none of them is as complete as we would like to see, they all do a reasonably good job of eliciting enough information about the nature of the invention to allow a patent professional to do a thorough patentability search. However, should the decision be made to draft an actual patent application, it is likely that some follow-up will be needed, and particularly if additional research or refinement of the invention subsequently takes place. You should always be prepared for requests for additional information, either in writing or by way of personal interview. Should your follow-up work on the invention produce anything that changes, however trivially, the character of the invention, you, yourself should take the initiative and pass this on to your patent attorney as soon as is practicable.

Where all of these standard forms fall short is in their omission of important information ancillary to the actual description of the invention. In Example 1, we find no place to enter any information about when the invention was first conceived, reduced to practice, and disclosed to others. Example 2, which is somewhat longer, not only fails to ask for this same information, but also requires the inventor to include "claims" to the invention, something that most inventors cannot be expected to do competently. It appears that the layout of Example 2 was derived from the format of actual patents without giving a lot of thought as to what additional information should be included, as well as what would better have been left out.

Example 3 suffers from a different sort of problem. Its entire first page is devoted to the collection of the kind of information better suited to an application for employment than an invention disclosure. This form is overly bureaucratic and has the unintended effect of discouraging anyone from filling it out because it is just plainly too much work. However, unlike the previous example, this form does have the virtue of directly asking for the date of conception. Interestingly, it falls down on one extremely important feature — unlike the shorter and simpler forms, it does not indicate that it needs to be signed and dated, let alone witnessed.

You will notice that none of the forms ask about prior sales activity, if any, prior publications, or prior public use.

Whatever their particular deficiencies, keep in mind that any patent professional would be much happier to receive a disclosure on any of the above forms than a page or two of scribbling and cryptic sketches, which is often all we are given, with the expectation that we will perform a prior art search or even turn it into a full-blown patent application. The sketchier and more unintelligible the disclosure, the more likely we are to misunderstand the invention, search the wrong prior art, and write an application that does not actually cover the invention. In order to do our job, we need you, the inventor, to do yours. The better and more complete the disclosure you give us, the stronger your ultimate patent will be.

Appendix. Exemplary invention disclosure document forms

Example 1

Inventors:
   Company Affiliation of Each Inventor:
   Addresses of Each Inventor (residence and work):
   Title:
Building and Testing Disclosure: 1. Date of conception; 2. Previous disclosure of the conception; 3. Substance of invention; 4. Impact of invention on field; 5. Tests; 6. Test results; 7. Names of individuals other than inventors to whom the concept was disclosed and dates of such disclosures; 8. Reference to relevant notebook passages by volume, date, and page numbers. Attached sketches, photos, printouts, and graphs as appropriate.

   1.   Date of conception:
   2.   Previous disclosure of the conception:
   3.   Substance of invention:

4. Impact of invention on field:
5. Tests:
6. Test results:
7. Names of individuals other than inventors to whom the concept was disclosed and dates of such disclosures:
8. Reference to relevant notebook passages by volume, date, and page numbers:

Inventor(s) signatures:

Date of signature(s):

The above confidential information is Witnessed and Understood by:

Signature of Witness 1:   Signature of Witness 2:

Date of Witness 1 Signature: Date of Witness 1 Signature:

Example 2

Inventors:
    Company Affiliation of Each Inventor:
    Addresses of Each Inventor (residence and work):
    Title:

1. Statement of Purpose:
2. Discussion of Prior Art:
3. Relevant Patent classification/keywords:
4. Advantages Over Prior Art:
5. General Statements of Invention:
6. Potential Patent Claims:
7. Reduction to Practice (includes figures as needed):
8. Relevant References:

Inventor(s) signatures:

Date of signature(s):

The above confidential information is Witnessed and Understood by:

Signature of Witness 1:   Signature of Witness 2:

Date of Witness 1 Signature: Date of Witness 1 Signature:

Example 3

Upon completion forward to:

Attorney: Docket Number:

Classification:

Product Category:

Inventor(s) (full name):

Company Affiliation of Each Inventor:

Addresses of Each Inventor (residence and work):

Employment Category of Each Inventor:

Salaried	Hourly	Retired
Supplemental	Agency	Consultant

Job Title of Each Inventor:

Organization of Each Inventor:

Phone and e-mail contact information for each inventor:

Citizenship of each inventor:

Supervisor name and contact information for each inventor:

Note: Notification is required when home residence information changes or upon leaving the organization.

Descriptive title of the invention:

1. What do you consider the new technology of the invention?
2. Identify the purpose/function of the new technology/technologies of the invention and the advantages over prior technologies.
3. Identify the closest technology, if any, of which you are aware. Provide copies if available. Identify the first dated record(s) of invention.
4. Date a working model, device, process, or composition was or will be completed.
5. If the invention will be released for publication, offered for sale or production identify action and date(s). Note: Many journals electronically publish months before paper publication.
6. If any, identify a government agreement, partnership, consortium, or other organization involved with conception or first building to the invention.
7. If the invention has been disclosed to nonorganization personnel, identify recipient and date.

Title:

Background:

    What is currently being used?

    Who currently produces it?

    Why is a new technology needed?

Specification of Invention:

How is it made?

Why is it made that way?

Drawings:

Other:

Part III

Relationships

chapter five

Ownership of intellectual property: employer rights to intellectual property

Ernest I. Gifford and Avery N. Goldstein
Gifford, Krass, Groh, Sprinkle, Anderson & Citkowski, P.C.

Contents

5.1   Introduction

Modern mythology describes the character of the inventor as being an individual toiling in the obscurity of a basement until the invention is patented and becomes a great convenience of life. In the process, the inventor is elevated to a household name. The reality of modern research and technology is far from the myth. Today professional researchers make the vast majority of innovations working in groups within complex corporate, academic, or governmental research organizations.

A basement inventor had no problem determining that ownership of the invention rested in his or her hands. In contrast, the ownership of an innovation made by an organizational inventor can rest in the organization, the inventor, or somewhere in between. The existence and substance of an employment contract is usually the determinative factor in assessing who has ownership rights in an innovation. This chapter discusses the details of employer rights to an innovation.

5.2   Employment agreements

The general rule is that, at least initially, intellectual property in the form of patent rights, copyrights, or trade secrets belongs to the creator of the property. Even trademark rights can, in a sense and in some situations, initially belong to the creator as opposed to the general rule that the first user of a trademark is the owner of trademark rights regardless of who creates the trademark.

Patent rights are embodied in a patent issued by the United States Patent and Trademark Office (USPTO) on an invention in the form of a new, useful, and nonobvious process, machine, manufacture, or composition of matter, any new and useful improvement thereof, or industrial design. Recently, methods of doing business have been added to the list of inventions that may be patentable. A patent does not confer the right to make the patented invention on the owner of the patent but instead grants the owner the right to prevent others from making, offering for sale, selling, or using the patented invention.

The inventor is the creator of the invention that forms the basis of the patent rights. Patent applications, at least in the U.S., have to be filed in the name of the inventor or the creator of the invention. The patent rights may be sold or assigned and often are the subject of transactions. The assignment of patent rights can be recorded in the Patent Office and the assignee then becomes the registered owner of any patent rights created by a patent issuing from the patent application.

Patent rights are considered to be personal property and therefore, unlike most of the law dealing with patents, which makes patents subject to federal law, laws of the individual states govern patent ownership. Like other forms of personal property, patent rights can be jointly owned. When two or more persons have contributed to the making of the invention, they are joint inventors because the joint inventors are treated as tenants in common, rather than as joint tenants. The distinction in joint ownership means that upon the death of one of the inventors, his or her share passes to his or her heirs rather than to the other joint inventors.

5.2.1   Basic employment agreement elements relating to intellectual property

Often the patent rights are assigned to the employer of the innovator. This is commonly the result of an employment agreement that is executed by the employee at the time that employment begins. A typical employment agreement requires the employee to perform the following acts:

- keep records of research undertaken as part of employment responsibilities;
- disclose any innovations to the employer or the employer designate, where innovations are normally broadly defined to include trade secrets, patents, copyrights, trademarks, and a catch-all of other confidential materials;
- assign all rights in innovations to the employer;
- participate, at employer expense, in the securing and defending of any innovations; and
- retain the confidentiality of the innovations.

The validity and interpretation of these agreements are determined in accordance with state law even though they involve a federally created right to protect inventions and the transfer of ownership of such rights. Like all agreements they require legal consideration, or generally a benefit to the employee for assigning his or her rights to his or her employer, but whether or not there has been a legally sufficient consideration to create a legally binding contract is a matter of state law. Generally there is no problem in finding the necessary consideration if the agreement is executed near the time of initial employment. The employment constitutes the legally necessary benefit to the employee. Courts have historically been reluctant to intervene in the matter and upset a contract struck between employer and employee to reallocate rights spelled out in an initial employment contract.

5.2.2 Common employment agreement issues

5.2.2.1 Employment agreement breach

It is sometimes the case that an employee refuses to assign or otherwise participate in securing rights to an innovation that was made under the terms of an employment contract. The common scenarios where this occurs are hostile employment termination, disablement, or death. The above provisions found in an employment contract are often the basis for an employer being granted the right to prosecute intellectual property rights independent of the inventor. A recalcitrant former employee certainly adds a measure of additional work by not participating in securing employer rights in violation of an employment agreement, but such efforts are rarely effective in preventing securement of rights. Additionally, a spiteful employee is exposed to legal liability for the breach of the employment contract terms.

The converse situation is where an employer fails to satisfy a contractual condition that is part of the employment agreement bargain. The obligations of an employee to assign his or her rights in inventions in light of an agreement breach by the employer are a matter of contract law. In the U.S., this means that this is a matter of state law. The obligation to assign rights per an employment agreement in the face of a contract breach by the employer (such as not paying wages, benefits, etc.) will depend on the severity of the breach. In other words, if the employee's original motivation for entering into the agreement has been frustrated, a court may hold that obligations on the part of the employee are obviated. However, in most instances the employer breach does not rise to the level of severity precluding employee compliance with the agreement. Alternatively, the employer can cure the breach by remedial actions. As such, an employee entering into an employment agreement at the time of employment should expect the terms of that agreement to be enforced.

5.2.2.2 Posthiring obligation to assign intellectual property

An employer seeking the execution of an employment agreement after the initial employment of the inventor may create a problem. Finding the consideration benefiting the employee necessary to support an agreement by the already employed employee to assign all of his or her inventions to her employer in those situations is difficult and sometimes impossible. When such agreements are found to be valid, the employer often has argued that the continued employment of the employee is sufficient consideration to support the agreement. This will not satisfy the requirement of legal consideration in most states. In this situation, where an employer is seeking to modify the employment agreement to extract additional value to intellectual property rights from an existing employee, a contract issue arises that requires consultation with an attorney about the law of the jurisdiction regarding contractual consideration.

5.3   Ownership of intellectual property

The ownership of intellectual property is dictated by a contract that allocates rights between an employee-creator and an employer. To avoid the entanglements detailed in this section, the creation of an employment agreement that addresses the allocation of rights is an easy solution. When there is no contract in place to determine the relative rights between and employee-innovator and the employer, the allocation of rights is dependent on the nature of the intellectual property.

5.3.1   Patents

5.3.1.1   Employment agreement scope

Often, even if a valid agreement has been entered into between the employee and the employer, there may be a question of whether or not a particular invention belongs to the employer rather than the employee. Generally, an employer can only claim rights to an invention directed to subject matter that has value or potential value with respect to the employer's business. More than that is considered overreaching since such contracts are often contracts of adhesion; namely, a contract in which one side, here the employer, holds the power and the employee has little or no say as to what will be the terms of the contract. Such contracts are strictly construed against the employer.

5.3.1.2   Obligation to assign absent an agreement

Where there is no written contract between the employee and the employer that defines the ownership of inventions created by the employee, the question will turn on the nature of the employee's employment. Is inventing a part of his or her expected duties? If research is part of his or her employment duties, then his or her inventions will belong to the employer as though a written contract had been executed agreeing to assign the inventions to the employer.

The question as to whether an employee has specific responsibilities to invent is critical in determining where rights to an invention lie absent an employment agreement. This becomes a factually difficult matter for an employer to prove since there is no agreement in place and the evidence turns on the daily activities and course of conduct of the employee. Employment responsibilities that include fabrication, assembly, design, and getting a production line up and running are likely to be construed as primarily nonresearch jobs, and therefore an employee having such work responsibilities may well be entitled to retain rights to her invention.

It is in the instance where an employer considers an employee not to be a researcher that employment agreement provided by the employer lacks the above requirements that the employee assign rights to any innovations to the employer. Typically, an employee working daily with a problem creates a solution that proves to have great value. In these most practical of innovations, the employer has the weakest claim to the innovation. Such an employee can be turned into a research employee and therefore have an inherent obligation to assign rights in any inventions through a change in work responsibilities to specifically include a research project or solving a particular problem. Since the burden rests with the employer to prove employment to invent absent an explicit contract, documentation to that effect and proper use of laboratory record-keeping mechanisms (see Chapter 7) are strong evidence in support of the work responsibilities of an employee working beyond the scope of explicit written obligation to assign.

5.3.1.2.1 Additional Factors in Determining Employer Rights Absent Agreement. There are a number of additional factors that are helpful in examining the rights an employer may have in an invention. These factors have arisen in disputes in various jurisdictions and should be considered representative. It is unlikely that all of these factors will be present in an actual factual setting and as such no one factor should be considered dispositive.

5.3.1.2.1.1 Past Conduct. The past conduct of the employee with previous inventions sheds some light on the verbal working agreement the employee had with the employer in the past. For instance, employee assignment coupled with a promotion and/or bonus for a past invention whether patented or retained as trade secret is suggestive of a bargain being struck between the employer and employee.

Of less evidentiary weight than the actions of the employee in question are the acts of like-situated employees upon bringing an invention to the attention of the employer. The understanding of other employees regarding the disposition of similar inventions is instructive in assessing the ownership of the disputed intellectual property. For instance, if another employee in exchange for a bonus payment assigned a similar invention, but the employer refused to pay a bonus for the assignment of the disputed intellectual property, then there is evidence that no agreement was reached between the employee and the employer. In this scenario, the employee would likely retain ownership perhaps subject to a shop right granted to the employer. The details of the shop right doctrine are discussed in Section 5.4.

5.3.1.2.1.2 Inventive Conception Date Relative to Employment Date. A factor of considerable weight is the timeline of the inventive process relative to the date of hire. If an employee can establish that an invention was conceived of before the date of employment, the invention is likely an act prior to the employment agreement and therefore not covered by the forward-looking language of that agreement. As a matter of contract law, attempting to contract for services already rendered is strongly discouraged absent clear payment for such services. In other words, while an employer can give additional value to an employee for prior conceived inventions (assuming the employee is not bound to assign to a previous employer), employment agreement language requiring assignment of all past and future inventions conceived lacks adequate compensation and is likely to be adjudicated critically.

5.3.1.2.1.3   Source of problem identification.   Whether the employer or employee identifies the source of the problem that the invention ultimately addresses can be used as a factor in determining with whom lies the impetus for the invention. This factor is of some importance because, unlike many other factors that focus on the relationship between employer and employee, problem identification is an initial step in the inventive process. Identification of the inadequacy of conventional technology is often a precedent to the conception of the invention. An employer who places the problem before its employee is laying a foundation for a beneficial invention. The context in which the problem is presented to the employee can provide evidence that the employee was given the task of working specifically on the problem. An employer who, after presenting a problem, makes available work time and or facilities to develop an invention has a strong argument for ownership rights even absent an agreement to assign rights in employee inventions.

5.3.1.2.4   Previous Statements.   The law is universal in considering previous statements about a subject as relevant evidence, at least as perception of the situation that is the subject of the statement. A statement made as part of a judicial proceeding or before a government organization such as the Patent Office can give rise to estoppel. This means that once a statement is on record, a contrary position cannot be asserted merely because it is now advantageous.

In the context of inventorship, an employee or employer who attributes the ownership of an invention to the other in the resolution of another matter can expect a critical adjudication if they later assert ownership themselves. Common disputes giving rise to previous statements that evidence attribution of ownership in the other party include patent cost payment, overtime payment for inventing activities, personal injury compensation related to the inventive act, and defense to a charge of patent infringement.

5.3.1.2.1.5   Payment of Patent Costs.   Paying for something is good evidence of belief in ownership. An employee may wrongly believe that they are owners in an invention and pay the costs associated with securing patent rights even though their employer asserts ownership in the invention. This can occur since an inventor absent input from an employer can initiate a patent application. The converse is not the case, an inventor who seeks payment of the costs associated with securing patent rights to an invention of their creation has, or should have, the reasonable expectation that when the employer pays patenting costs, they are doing so with the expectation of some form of ownership interest. An employer who refuses to patent an invention at employer expense is a more ambiguous situation since an employer may feel the invention is better suited for maintenance as a trade secret, or believe the invention is not valuable. In either of these circumstances an inventor may still assert an ownership interest in the invention.

5.3.1.2.1.6 Position within the Organization. The determination if someone has an inherent obligation to assign rights in an invention to an employer is often dictated by the position he or she holds within the organization. The most straightforward case is a person hired as a researcher where the obligation to assign is recognized as part of the employee service rendered. As a result, there are few circumstances of conflict as to invention ownership between a researcher and the employer.

With individuals who are other than research employees, the obligation is less clear. In these instances a review of the other factors becomes appropriate to determine if the employee–employer relationship is consistent with a transfer of rights resting with the inventor to the employer.

In the case of an independent contractor-supplied worker, the employer has not entered into an employment contract with the worker, but instead with the contracting company that employs the worker. In such a circumstance, absent an agreement signed by the worker with the contracting company, the claim of the employer for an assignment to an invention created by such a worker is tenuous. This is true even if the worker is a researcher, since the worker is not paid directly by the employer, nor does the employer employ the worker. The worker obligation to assign, if it exists, requires assignment to the contracting company, and not the employer. The immediate response to this situation is that practically the contracting company in order to continue to provide workers to the employer will reassign any rights to the employer. While this reassignment is likely to happen, this scenario still has negative implications on patent prosecution in situations where the employer is prosecuting patents in the U.S. and asserting common ownership at the time of invention in order to avoid the employer's own patent portfolio being leveled against the invention. In the scenario of a reassignment from contracting company to the employer, no assertion of common ownership at the time of invention can be made, thereby jeopardizing the patentability of the invention.

A guest worker can be considered an independent contractor-supplied worker for the purposes of employer ownership claim in inventions. In this instance, the contracting company is the individual himself or herself.

An officer or a director of an organization often does not have a written agreement in place that dictates the disposition of an innovation he or she makes that is relevant to the organization. In this situation, the organization has a strong argument that the officer or director has a fiduciary duty to transfer the rights to the invention. This obligation is derived from the state law defining the incorporation of the organization and implies an inherent duty as part of their position of responsibility to act in the best interests of the organization.

5.3.1.2.1.7   Royalty Payments.   An employer who pays a royalty payment to an inventor has created through course of conduct a situation that suggests that the inventor has retained ownership of the intellectual property. A conflict between inventor and employer in this context often results when there is a disruption in the royalty payment scheme such as a demand for a change in royalty terms, a better offer to the inventor by a third party, or a charge that the employer is infringing a third party patent by practice of the intellectual property. This is true not only of patents but also trade secrets.

5.3.1.2.1.8   Employer Response to Invention Disclosure.   An employer who is disinterested in the creative work of an employee has created through course of conduct a situation where later assertions of ownership will be adjudicated critically. An employer who fails to act to protect an employee invention may be exposed to liability in situations where the employee is entitled to a bonus, or percentage of the royalties derived from patented invention. In the instance where an employer has failed to act to secure patent rights, an employee upon showing that the innovation was patentable and what the value would likely have been may well receive a judicial award as compensation for lost royalty payments. In recognition of this liability exposure, it is common for an employer to have in place a process by which an invention the employer is disinterested in protecting is reassigned to the inventor for him or her to exploit. Terms associated with reassignment of invention rights to the inventor vary widely from the employer granting only a limited term license in exchange for receiving a royalty and reimbursement of costs incurred to reassignment of all employer encumberances.

   5.3.1.2.2   Reviewing Employment Agreements for Assignment of Intellectual Property.   With this situation in mind, it is important that an employer review employment contracts of all employees and workers to determine if assignment of innovation language is sensible, and to also create a mechanism by which such innovations are communicated to the employer. The most active form of obtaining innovations from employees involves a bounty system where the submitting employee is given a fixed reward or a percentage of the saving/profits resulting from the implementation of the submission.

5.3.2   Trade secrets

State laws largely govern trade secret law. Most states have adopted the Uniform Trade Secret Act which generally defines a trade secret as "consisting of any formula, pattern, physical device, idea, process or compilation of information that both:

- provides the owner of the information with a competitive advantage in the marketplace; and
- is treated in a way that can reasonably be expected to prevent the public or competitors from learning about it, absent improper acquisition or theft."

A formula such as the one for Coca-Cola® is an example of a valuable trade secret. The rights to a trade secret only exist as long as the formula, drawings, business methods, and even inventions are kept secret. Protection is afforded against acquisition by wrongful means. Acquiring the trade secret in a lawful manner is not actionable.

The employer rather than the employee generally owns trade secrets. One of the requirements of creating a trade secret is that reasonable efforts be made to keep it secret. These efforts usually include an express or an implied agreement by employees in a position to know the secret that they do their best not to disclose the secret to outsiders. Such an agreement is an acknowledgement by the employee that the trade secret, regardless of its source, belongs to the employer.

Controversies involving the ownership of trade secrets often arise when the employee leaves the employer. Although there are cases in which the courts have held that the trade secrets belong to the departing employee, either because the employee created the trade secrets without an obligation to do so, or because she brought the trade secrets with her when she entered into the employment relationship, most of these situations are decided on the issue of whether or not there was an implied or an express agreement setting forth the ownership of the employer's trade secrets and whether or not the items in question are in fact trade secrets.

Again, although in most instances, at least initially, the creator of the item that becomes a trade secret is the original owner of the property rights, the item ownership is transferred to the employer soon after creation either by an implied or an express agreement.

5.3.3 Copyrights

Copyright is a form of protection provided by the laws of the U.S. to the authors of "original works of authorship," including literary, dramatic, musical, artistic, and certain other intellectual and artistic works. This protection is available to both published and unpublished works. Unlike other forms of intellectual property such as patent rights and trade secrets, federal law largely determines the initial ownership of a copyright.

At least some of the rights afforded to copyright owners come into existence when the "pencil leaves the paper" or in other words at the time of creation. Additional rights are obtained when an application is filed in the U.S. Copyright Office and a registration is obtained. The application is filed in the name of the "author" who can be either the creator or the employer of the creator where the work is considered to be a "work for hire." An assignment of the copyright must be in writing and can be filed either with the application or later.

The author is the employer where an employee does the work and the employee has agreed either impliedly or by a written employment agreement that copyrights arising out of the employment will belong to the employer. If there has been an assignment of the copyright prior to registration, the application is filed in the name of the author but filing the assignment with the application will record the assignee as the owner of the registration.

One who commissions certain types of works is also the "author" for purposes of filing a copyright application and obtaining the copyright registration of the commissioned works provided the works are in a form specified by the copyright statute.

In the case of works made for hire, the employer and not the employee is considered to be the author. Section 101 of the copyright law defines a "work made for hire" as:

(1) a work prepared by an employee within the scope of his or her employment; or
(2) a work specially ordered or commissioned for use as:
   • a contribution to a collective work
   • a part of a motion picture or other audiovisual work
   • a translation
   • a supplementary work
   • a compilation
   • an instructional text
   • a test
   • answer material for a test
   • an atlas
if the parties expressly agree in a written instrument signed by them that the work shall be considered a work made for hire ....

5.3.4 Trademarks

A trademark is a name, slogan, logo, or the like that is used with a product or service, which indicates the owner of the mark as the source of the goods or services (service mark) and distinguishes the goods or services of the owner competitors. Chevrolet is a trademark as is the MGM lion and the NBC chimes. Because trademarks and service marks serve to indicate the source of goods and services, it follows that they are owned by the business entity, which is the source of the goods and/or services, rather than the creator of the mark. Ownership of trademarks and service marks is assigned by use rather than by creation.

Registration, either state or federal, is not required to obtain the right to use a mark or to prevent others from using the unregistered mark.

Use, rather than creation of a trademark or service mark defines ownership between the source of the goods or services and those who use a confusingly similar mark for related goods. However, the creator may have contract rights against the user of the mark. If the mark were developed for, or offered to, the user, then contract rights, either implied or express, may define the rights between the user of the trademark and its creator.

5.4   Shop rights and assignments

If there is no written agreement, and the employee is not in a position in which he or she is expected to invent, the employer may still have limited rights to his or her inventions. An employer has a "shop right" in the inventions of its employees where the invention is made on the employer's time with the aid of the employer's resources. A shop right, however, only permits the employer to use the invention in its business and does not permit the employer to sell products utilizing the invention or to license others to do so. A shop right can exist in an idea that fails to meet the requirements of an invention.

5.4.1   Employer time and facilities

An invention developed during employee work hours and/or using employer facilities absent an agreement to assign rights is the most common scenario giving rise to a shop right. The origin of this doctrine is instructive in its application in a specific situation where the facts are less than clear. The doctrine of shop rights has an origin in English and other common law jurisdictions as an equitable remedy. The basis for the shop right is that there is no legal obligation to enforce against the employee who invented with the benefit of employer facilities, yet it is inherently unfair for the employer to have lost the benefit of employee labor and or the lease payment for the use of facilities; and further be obligated to pay the employee a royalty on the use of the invention or to be denied the use of the employee invention. Shop right litigation most often occurs when the employee is offered a lucrative license and employment with another firm, leaving the employer at a competitive disadvantage through the employee use of employer time and/or facilities to invent.

5.4.2 Shop right limits

There is, among other origins, an equitable nature to a shop right correcting an unfair outcome. This means that the rights conferred to the employer are quite limited. The employer does not have the ability to grant a sub-license to a shop right or otherwise transfer the right, independent of complete sale of the employer assets. Clever attempts to disguise a transfer of rights from employer to a third party generally are viewed by the courts with skepticism.

The term of a shop right varies with the status of the invention. When a shop right exists in a patented invention, the term is typically the life of the enforceable patent, absent an agreement between the parties for a shorter duration. Thereafter, the employer, like any other member of the public is free to exploit the invention without accounting to the inventor. A shop right in a trade secret may survive indefinitely so long as that status is preserved. As the shop right is born out of an effort to provide fairness, judges are free to provide less than these terms of shop rights, based on the conduct of the employer and employee. This ability to override legal defaults with an agreement extends to the existence of the shop right itself. Should the parties agree to a royalty, then the invention is subject to a license and the shop right is extinguished.

5.4.3 Functions outside traditional employment relationship

The term "shop right" is appropriate only in an employment relationship; however, the equitable features of this doctrine may well be applied to nonemployment scenarios that are becoming more prevalent in the workplace. Rather than delve into the divergent views and the resulting ambiguity associated with an extension of a shop right doctrine to a nontraditional relationship between a worker and a benefactor of the labor, the matter is best addressed by a prior written agreement establishing the disposition of invention rights. A review of the relevant law in the U.S. is available to the interested reader in the Additional Readings.

5.5 Summary

In the absence of a contract, either implied or express, patents, trade secrets, and copyrights initially belong to the creator. A contract requiring an employee to assign those rights to an employer can be implied from the nature of the employment duties. In the matter of copyrights and trade secrets original ownership can be in the employer and not the creator where property is a work for hire or where the employer's resources have been used to create a trade secret as a part of the employee's duties.

Trademarks are owned by the source of the product or services of the trademark user rather than the creator. However, the creator may have contract rights that provide the creator with rights to compensation for the user's use of the mark.

Ownership issues represent a complex and highly divisive issue within an organization. The prospect of significant financial gain by the holder of invention rights only adds to the acrimony. By addressing the allocation of invention rights at the outset of an employment relationship there are clear expectations as the disposition of invention rights, and therefore ownership issues are more readily managed.

A written contract is the cornerstone of orderly invention right allocation. The parties are free to allocate rights as they see fit, with the exception of contract terms that are offensive on public policy grounds. Through resort to a written invention allocation agreement, the parties are governed by an agreement of their choosing and not the legal defaults relied upon by courts in the absence of such an agreement. This is the single most valuable procedure that can be put in place to avoid subsequent internal conflict.

A written contract governing disposition of an invention can also extend to other innovation aspects addressed in this chapter. For instance, an employee may wish to use employer facilities to develop an invention beyond the scope of their employment. Most often an employer will grant use with little or no expectation of rights in the invention, provided the employee pays for materials consumed, tinkers outside of work hours, and signs a waiver as to injury liability while working on the project. Documentation of project supply purchases, concepts, and results in a bound and page-numbered notebook or digital equivalent affords numerous advantages in providing evidence of inventorship and ownership interests. It is often the case that avoiding the minor inconvenience of preparing an invention allocation agreement as an initial matter results in subsequent complications of considerable proportions.

5.6   Exercises

5.6.1   Exercise 1

Ben Gay is employed by the Aches & Pains Fitness Club. The employment agreement that he signed when he started work for the club is silent as to the ownership of inventions made while he is employed by the club. His job from the start of his employment has been to keep the pool area clean. As a part of his duties he sweeps the pool locker rooms each day. While on his break after completing his sweeping task, he was looking over the broom that he was using and a significant improvement in the design of the broom suddenly came to him. The design that he envisioned would make the broom much easier to push. He immediately began work on his new design. It took about 12 hours to complete the project, with about half of the time being at work and the other half at home.

Who owns the invention? Ben or his employer?

Since the employment agreement that Ben signed is silent on the issue of ownership of any inventions made by Ben, there is no express agreement to assign his inventions to the club. There is nothing to indicate that there is an implied agreement. Ben was not hired to invent and inventions would not be expected to come out of his duties at the club. The broom is not a product that the club ordinarily makes and sells. Ben owns the invention.

The most that an employer might have in the way of rights to practice the invention would be a shop right. A shop right in favor of the club might arise because at least a part of the development took place on company time and using company facilities. A shop right is defined under state law and in most states a shop right gives the employer the right to use the invention in its own business but not the right to sell products incorporating the invention or the right to prevent others from doing so. Those are rights exclusive to the patent owner. If the club has a shop right, then it could make or have made brooms of the design for use by the club but it would not have the right to patent the broom or to sell improved brooms to others.

5.6.2   Exercise 2

Glen Hobbit has an idea for a computer game but he does not know how to create the software necessary to produce a game from his idea. He hires the Lord Computer Design Company to use his idea to produce the software for a computer game based upon his idea. He agrees to pay the company $12,000 upon completion of the software for the game. The company assigns the task to its employee, Jack Lord the son of the founder of the company, to complete the project. Nine months later and after a lot of anxious contacts by Glen the project is finally completed.

Who is the owner of copyright for the game? Glen Hobbit, Jack Lord, or the Lord Computer Design Company?

In the absence of an agreement to assign the copyright the Lord Computer Design Company is the owner of the copyright. All Glen had was an idea. Ideas are not copyrightable. Only the expression of the idea can be copyrighted and Jack Lord is the creator of the expression of the idea in the form of the software to play the game. He is, however, an employee of the Lord Computer Design Company and his creations made as a part of his job are copyrightable by, and belong to, his employer under the "work for hire" doctrine.

Additional reading

Chisolm, D.S., Chisom on Patents, 89th rel., Matthew Bender & Co., New York, 2003, pp. 22–55.

Mills, J.G. III, Highley, R.C., and Reiley, D.C. III, *Patent Law Fundamentals*, 2003 rev., Eagan, Minnesota, 2003.

chapter six

Inventorship

Angela M. Davison
Ross Controls

Contents

6.1 Introduction

From classical inventors like Johannes Gutenberg, Benjamin Franklin, and Thomas Edison to modern marvels including Dean Kamen and Steve Jobs, inventors have a unique ability to shape the course of human history. Yet not all inventions have to be grand to impact the way people live. Most of the appliances and gadgets in public use are not earth-shattering innovations, but instead are everyday comforts and conveniences that make our lives a little easier to manage.

Most scientists and engineers have thought of ways that currently existing devices and ideas can be improved. In fact, most inventions submitted to the United States Patent and Trademark Office (USPTO) can be classified as improvements to already existing devices and technologies. While most scientists and engineers have an intuitive idea about who an inventor is, they are often unaware of how to define the status of inventor from a legal standpoint. Scientists and engineers, by the nature of their employment, may find themselves in the position of applying for a patent. As inventors, they need to understand how an inventor is legally defined in order to avoid having their patent invalidated because inventorship was improperly listed. This chapter will adequately educate scientists and engineers on the subject of legal inventorship.

6.2   Why it is important to determine proper inventorship

Unlike most industrialized nations in the world, the patent system in the U.S. is characterized as a "first-to-invent" system. A first-to-invent patent system is one where a patent can be granted only to those who are the first to invent the subject matter of the patent. In comparison, most other industrialized nations use a "first-to-file" patent system, where a patent is granted to the individual or entity who files a patent application first, irrespective of who was actually the first to conceive the invention.

The policy underlying the U.S. first-to-invent system is the protection of the true inventor and the protection of the public. The first-to-file system protects the interest of the true inventor by preventing others from reaping the benefits of the inventor's work. If the true inventor's efforts are not rewarded with a patent grant, there is little incentive for inventors to design and develop the methods and devices that enhance our lives. The public's interest is protected because the power to exclude others from making or selling the invention is limited to the owner of the patent.

Clearly, in first-to-invent patent systems, the determination of the true inventor is important because only the true inventor can receive a patent. Yet determining exactly who can be an inventor is not quite as clear.

6.3   Determination of inventorship

There are many legal definitions of the term "inventor." One definition states simply that an inventor is one who conceived of the subject matter at issue. Another definition states that an inventor is one who conceives of an idea, and has this idea captured in a claim of a patent. Still another definition states that an inventor is one who conceives the solution to a problem and the means to attain the desired solution to the problem, where these means constitute the subject matter of the invention. Despite the plethora of definitions, none are particularly instructive in helping one determine if one is an inventor. However, there is a two-step process that might ease the difficulty of making an inventorship determination: (1) identify those who conceived the subject matter at issue; and (2) confirm that this subject matter was recited in a patent claim. Both of these steps are discussed below.

6.3.1   Conception of the subject matter at issue

Conception is the mental development and disclosure by an inventor of an idea for an invention [1]. An invention is conceived when the inventor thinks of a solution to a problem. For this solution to be classified as an invention, it needs to be developed enough to allow a person of ordinary skill in the art to practice the invention [1]. A person of ordinary skill in the art is someone who has an ordinary level of proficiency in the particular technical area in which the invention is made [1]. For example, if the technical area in which an invention falls is plumbing, a person of ordinary skill in the art is a plumber, not a Ph.D. civil engineering professor.

Clearly, if you worked completely by yourself and did not seek any input from others at any stage during the conception and development process, then the question of who conceived the invention is quite easy to discern. Yet in most cases, the first exposure a scientist or an engineer will have with the patent system is in an employment context as part of a development or research team. The scenario where more than one person conceives of the invention is discussed in detail in Section 6.6.

6.3.2   Recitation of the subject matter in a patent claim

It is not enough to conceive an idea to be considered an inventor. Rather, one must also verify that the idea conceived was captured in a patent claim. Often, especially in the corporate context, when submitting an invention disclosure form to the corporate management to obtain approval to apply for a patent, one will be asked to list all of the inventors. Technically, such a list cannot be provided accurately until after the patent application is written. Thus, when completing the invention disclosure form, one should list all of the individuals one believes contributed to the invention as inventors. Then, when the application is reviewed before it is submitted to the USPTO, the claims should be examined. While examining each claim, identify those who contributed the subject matter of the claim. These individual(s) will be the inventor(s) of that particular claim. Once all the claims are reviewed, verify that all the different inventors identified as contributors to the patent claims are listed on the invention disclosure form.

6.4   Inventorship and ownership

It is quite easy to confuse ownership of a patent with the inventorship of the subject matter in a patent. However, the two concepts functionally are worlds apart. Perhaps the most fundamental difference between ownership and inventorship is that the ownership of a patent can change at any time. A patent, just like any other piece of property, can be bought, sold, licensed, or held in trust. While the owner of a patent can change as a result of some type of transaction, the status of inventor cannot be changed by contract. The only way an inventor can be established as such is by meeting the criteria discussed in Section 2.2. Thus, one cannot sign an agreement to become an inventor.

The likely reason that ownership and inventorship are confused is because an inventor, by default, is deemed to be the owner of his patent [2]. However, there are three exceptions to this general rule, all of which occur in an employment context:

1. An employer owns the invention if the employer and the employee have a contract to that effect.
2. An employer owns the invention if the employee was hired to, or later instructed to, solve a specific problem in an area.
3. An employer, under certain circumstances, may have a "shop right" to use an employee's invention [3].

Each of these exceptions is discussed in detail below.

6.4.1   Employer–Employee contract

This exception to the general rule that an inventor is the owner of his invention tends to be the most common exception that a scientist or an engineer will encounter. Most scientists and engineers who are not self-employed are required to sign a contract with their employer upon hiring including a clause stating that all inventions developed by the employee belong to the employer. In other words, the employee, in exchange for the benefit of having employment, agrees to assign ownership of his inventions over to his employer.

EXAMPLE 1: Suppose Amanda, a Ph.D. chemist, was hired by ABC Pharmaceutical Company to perform general research and development. Upon hiring, ABC Pharmaceutical required Amanda to sign a contract outlining the terms of her employment. One of the clauses in this contract stated that

Amanda would be required to relinquish all of her rights in any intellectual property that she develops as a result of her employment. Amanda signed this contract and commenced her employment. After 5 years, Amanda successfully developed a drug to treat recurring neuralgia and applied for a patent. Amanda's invention, patent application, and any patents resulting from her invention belong to ABC Pharmaceutical because of the employment contract Amanda signed with ABC Pharmaceutical.

6.4.2   Employed to invent doctrine

The second exception to the general rule that the inventor owns his invention is when an employee was hired to, or later instructed to, solve a specific problem in an area [3]. This exception is referred to as the "employed to invent doctrine." This doctrine is applicable when an employee does not sign an agreement to turn over his inventions to his employer, but the employee is expressly hired to, or instructed to, invent a solution to a specific problem.

Suppose that in Example 1 Amanda's employment contract did not require her to relinquish all of her rights in any intellectual property she developed during the course of her employment. Also suppose that Amanda was the world's leading expert on the use of drug therapies to treat neurological disorders. Further suppose that ABC Pharmaceutical specifically hired Amanda to develop a drug to treat neuralgia in HIV patients, and made this intention clear to her. Five years later, when Amanda successfully developed the drug and applied for a patent on the drug, her invention, patent application, and issued patent will belong to ABC Pharmaceutical because she was expressly hired to solve a specific problem.

Most situations triggering the employed to invent doctrine are not as cut-and-dry as the above example. To assist both employers and employees in determining whether the employed to invent doctrine is triggered, courts have developed ten factors to aid in the determination [3]:

1. Were there any previous assignments of an invention or a patent to the employer by the same employee?
2. Is it customary practice within the company for similarly situated employees to assign their inventions and patents over to the employer?
3. Was the invention conceived during the employee's period of employment with the employer?
4. Who originally posed the problem solved by the invention, the employer or the employee?
5. Does the employee have the authority within the company to determine to whom to give a problem to be solved?
6. How important is the invention to the employer's business?

7. Has the employer had a previous inconsistent position on ownership of an invention?
8. Is there an agreement between the employer and employee to pay
   royalties to the employee?
9. Who paid for the patent application — the employer or the employee?
10. Was there an initial absence of interest on the employer's part
    when the employee first disclosed the idea?

An example to illustrate the application of these ten factors is included
below.

EXAMPLE 2: Patrick has been a mechanical engineer for Cofield's Chassis
and Engine for the entire 23 years of the shop's existence. Cofield's is a
smaller operation, currently employing about twenty people. Cofield's spe-
cializes in customizing and retrofitting engines in classic muscle cars. Mike
Cofield, the owner of the shop, hired Patrick on the spot after seeing
Patrick's completely rebuilt 1970 GTO. Patrick installed a 454 big block
Chevy engine, and designed and built his own custom intake valves, ex-
haust valves, and pistons that resulted in a 15% increase in low-end torque.
Mike admired Patrick's ability to design and fabricate completely new
components, and immediately knew that Patrick's talent would come in
handy.

   In the course of completing various projects for Cofield's over the years,
Patrick developed several improvements to existing valve technology.
Three of these improvements were patented. Cofield's paid for the patent
prosecution process for all three patents. Mike and Patrick never discussed
who owned the patents, but Mike assumed that Cofield's did because it
paid the prosecution fees.

   Mike was grateful to have Patrick as an employee, not only because of
his talent but also because Patrick's skill was widely known. One of
Patrick's valves was featured in a popular automotive magazine, and directly resulted in a substantial increase in business. Mike licensed the featured valve technology to two valve manufacturers, and the revenue
generated by the licensing agreements accounted for one quarter of
Cofield's gross profits annually. As a reward and an incentive to Patrick,
Mike decided to give Patrick 10% of the royalties earned on the licensing
agreements.

   Mike also employed Mark, another mechanical engineer. Mark is Mike's
brother-in-law and has been working for Mike for 13 years. Mark's specialty
was in heating, ventilation, and air-conditioning (HVAC) systems. Mark
was recently awarded a patent on a component he created to modernize an
HVAC system in a 1967 Mustang GT Fastback. Mike believed that Cofield's
owned this patent, but he and Mark got into a terrible argument over the
issue of ownership. Mike ended up conceding ownership of the patent to

Mark just to keep family relations peaceful. Mark bragged to the rest of his coworkers that he was now the owner of a patent, and that it hopefully would generate enough revenue so that he could retire in 5 years.

Shortly after Mark received his patent, Mike suggested to Patrick that he should apply for a patent on his latest valve innovation. Both Mike and Patrick were very excited about the invention, and both had reason to believe that the new valve would be extremely successful in the market-place. Patrick prepared a detailed invention disclosure and provided it a patent attorney, who billed his fee to Cofield's.

One afternoon, a few weeks after Patrick's patent application was filed, Patrick and Mike were having lunch together. Patrick mentioned that he was looking forward to being able to license his latest invention, and that he was thankful that Mike now has the policy of letting inventors keep their inventions. Mike was puzzled by this comment and asked Patrick why he thought there was a policy to let the employees keep their inventions. Patrick responded by pointing out that Mark was allowed to keep his patent. Mike said that the only reason Mark was allowed to keep his patent was because Mark is a member of Mike's family. Patrick wondered whether there is any way that he could possibly be considered the owner of his patent and made an appointment to see an attorney the next day. Who owns Patrick's patent — Cofield's or Patrick?

The most efficient way to determine the ownership of Patrick's patent is to apply the ten factors noted above.

1. The first factor weighs in favor of Cofield's because it owned all three of Patrick's previous patents, and Patrick never objected to this arrangement.
2. The second factor weighs somewhat in Patrick's favor. It is true that Mark, an employee in a similar situation, was allowed to keep his patent. However, the rationale behind allowing Mark to keep his patent was not based upon workplace considerations. Moreover, it cannot be said that one instance of allowing an employee to own his patent rises to the level of a customary practice. Thus, the factor technically weighs in Patrick's favor, but it is not a strong indicator.
3. The third factor clearly weighs in favor of Cofield's because all of Patrick's inventions were conceived while Patrick was on the job.
4. The fourth factor does not weigh in favor of either party because there are no facts in Example 2 indicating that either party origin-ated any of the problems solved by Patrick's invention.
5. Similarly, the fifth factor also does not weigh in favor for either party.
6. The sixth factor weighs in favor of Cofield's because one of Patrick's inventions constitutes a quarter of Cofield's' gross profits, and the pending application is also expected to be profitable.

7. Clearly, the seventh factor weighs in Patrick's favor because Mark was allowed to keep his patent.
8. The eighth factor weighs in favor of Cofield's. Even though Patrick was awarded with a percentage of royalties on one of his previous patents, Cofield's did not make any agreement with Patrick to give him royalties on his latest invention.
9. The ninth factor supports the ownership claim by Cofield's because it paid all of the patent procurement expenses for each of Patrick's patents.
10. Finally, the tenth factor does not weigh in favor of either party because the facts recited in Example 2 do not indicate the level of Mike's interest when Patrick created his inventions.

In sum, factors 2 and 7 weigh in favor of Patrick owning his patent, and factors 1, 3, 6, 8, and 9 favor ownership of the patent by Cofield's. Thus, a court would likely determine that Cofield's would be the owner of Patrick's patent.

6.4.3   An employer's shop rights

The final exception to the general rule that the inventor owns his invention is when an employee uses his employer's resources to conceive or build his invention [3]. In this case, the employee must give his employer a nonexclusive, royalty-free, nontransferable license to make use of the invention [3]. Granting someone a nonexclusive license means that the entity acquiring the license does not have the exclusive right to make the patented invention. Rather, the owner of the patent can license the patented invention to others as well. Making a license royalty-free means that the entity acquiring the license is not charged a fee for making the patented invention. Finally, a nontransferable license is where the entity acquiring the license cannot, in turn, sell or license his interest in the original license to some other entity.

Technically, a shop right is not an ownership interest in a patent. Thus, a company having a shop right does not have the right to buy or sell the underlying patent or invention. Instead, those ownership rights belong to the inventor. Creating a shop right simply reimburses a company for the employee's use of the company's resources while allowing the employee to retain ownership of his invention. A company having a shop right may make use of the invention until the patent expires.

The determination of whether a shop right is created can be complicated because it is difficult to determine the threshold amount of an employee's usage of the employer's time and resources that will require the employee to grant the employer a shop right. Courts have created three factors to aid with this determination [3]:

1. Did the employee develop the idea during work hours for which he is paid?
2. Did the employee use the facilities of his employer to produce an embodiment of the invention?
3. Did the employee introduce or allow the employer to introduce that embodiment into the employer's productive facilities?

Of these three factors, courts generally consider the second factor as being the most critical in a shop right determination. For example, courts have granted a shop right to employers even when the employee developed and built his invention on his own time, with his own equipment, but used his employer's facilities only to test his invention. Hence, if an inventor does not want his employer to have a shop right in his invention, it is critical that the inventor not use any of his employer's resources whatsoever.

EXAMPLE 3: Pam works full-time as a data entry clerk for XYZ Credit during the day. At night, Pam is taking courses at her local university to receive a bachelor's degree in computer programming. Pam has enjoyed computer programming ever since she went to a computer camp when she was 10 years old. It seems that no matter what computer software application she is running, Pam can always seem to find ways to improve upon it.

Pam has three children and cannot afford to have them in day care 5 days every week. To defray day care costs, Pam made arrangements with XYZ Credit to work from her home on Mondays, Wednesdays, and Fridays. This way the children only need to be in day care 2 days each week. On the days she works at home, she can watch the children while she works. XYZ Credit has supplied Pam with a laptop computer and a fax/copier so she can work at home more efficiently. XYZ Credit also pays for the Internet service at Pam's house.

One day while working from home, Pam's Internet connection to XYZ Credit's employee portal seemed uncommonly slow. Pam was able to determine that, despite having a functioning firewall, spammers had gained access to her computer and were running programs that drained her system's resources. Naturally, Pam was quite irritated and wanted to devise a way for her computer to detect the presence of spammers and deny them access to her system.

Every night after class, when the children were asleep, Pam worked on developing a computer program that detected spammers and prevented them from accessing her system. About half the time, Pam developed the program on her laptop, but the other half of the time Pam worked from her personal desktop computer in her living room so she could watch TV while she worked. Once she finished developing and testing her computer program and was satisfied with the results, she installed it on both her laptop and her desktop computers.

The next time Pam worked from home using her laptop, she noticed a marked improvement in the speed of her Internet access. As a result of the improved performance of her laptop, it took her a far shorter time to complete her work quota. This efficiency gave her more time to spend with the children and allowed her to take a nap before class.

After a few weeks of using the program, a classmate and coworker, Valerie, noticed that Pam looked more rested than usual in class one night and commented on it. Pam told Valerie how her computer program allowed her to be more efficient, thereby giving her time to take a nap before class. Valerie asked for a copy of her program, and Pam gave it to her.

After trying Pam's program several times, Valerie was quite pleased with the increased efficiency. At work, Valerie told everyone about Pam's terrific program. Soon, many of Pam's coworkers were requesting copies of her program. When Pam's manager Lisa got wind of Pam's program, Lisa also asked for a copy. After trying the program out for several days, Lisa decided that everyone in the office should have a copy. Within 2 weeks, every computer at XYZ Credit had a copy of Pam's program.

Pam, noticing the welcome reception her program had received, decided that it would be a good idea to apply for a patent. Once Pam received her patent, which included claims directed to further improvements in the program's efficiency, Pam met with Lisa and offered to license the program to XYZ Credit at a reduced rate. However, Lisa believed that XYZ Credit should be able to use the program for free. Does XYZ Credit have a shop right in Pam's program?

After analyzing the three factors noted above, it appears that XYZ Credit does have a shop right in Pam's program.

1. The first factor weighs in Pam's favor because she developed the program on her own time after work; thus, she was not paid to develop the program.
2. The second factor weighs in XYZ Credit's favor because Pam occasionally did use the laptop provided by XYZ Credit during the development of the program.
3. The third factor also weighs in XYZ Credit's favor because Pam allowed Lisa to distribute her program throughout the company.

Hence, because the overall tally of the factors weighs in XYZ Credit's favor, XYZ Credit does have a shop right to use the program. However, remember that just because XYZ Credit has a shop right to use the program, it does not mean that XYZ Credit owns the invention or the patent. The invention and the patent remain Pam's property, and she is free to sell or license the invention as she sees fit. It is important to note that Pam could have avoided creating a shop right if she had used her own computer exclusively in developing the program and if she did not allow Lisa to distribute the program at work.

EXAMPLE 4: Peter works for Master Impressions, an interior design firm. Peter creates custom-crafted stained-glass windows for the firm's clients. Peter has been creating windows and other artistic works out of stained glass for nearly 15 years and has a complete workshop in his garage. Peter has a master's degree in materials engineering and briefly worked as a ceramics engineer for Indianapolis Ceramics before realizing that he could make more money and have more fun turning his hobby into his profession.

When creating windows for the firm's customers, Peter would often be frustrated because the only available means to create boundaries within a stained-glass design was by the use of solder, which is dark gray or black in hue. Peter believed that having only one color of solder with which to make boundaries limited the range and style of aesthetically pleasing stained-glass designs. Peter wanted the ability to create boundaries using a material that had virtually unlimited color options. After work one day, Peter drew upon his ceramics background and came up with a couple of ideas for composite ceramic materials that might be able to function as boundaries in stained-glass designs. However, Peter did not have the facilities to create or test his ceramic composites.

Shortly after Peter came up with his ceramic composite ideas, he called his former supervisor, Bryan, at Indianapolis Ceramics and asked if he might be able to work on developing his ceramic composite ideas after work hours. Bryan agreed to let Peter use the facilities because Bryan figured that in the event Peter does develop a suitable ceramic composite, Peter would likely be able to direct some business Bryan's way.

In about 20 weeks, Peter was able to develop a suitable ceramic composite to create boundaries in any possible color in a stained-glass piece. Peter produced enough of the composite for his immediate needs at work. As soon as he was able, Peter filed a patent application for his composite, and began to seek licensing agreements with stained-glass material suppliers to make his invention. In the mean time, Peter remained employed with Master Impressions, and used the composite within several of his works for Master Impressions' clients. Peter's colleagues at Master Impressions greatly admired his new composite, but because Peter was the only stained-glass designer employed at the firm, no one else at work used the composite.

Within a matter of months, Peter's licensing efforts generated enough revenue for him to retire from his job at Master Impressions. Peter bought a small place in the Florida Keys and intended to spend the rest of his days fishing. One day while out on his fishing boat, Peter received a call from Joy, the general manager of Master Impressions. Joy wanted to know if her new stained-glass designer could make Peter's composite for use in stained-glass works for the firm's customers without paying Peter a royalty. Joy said that because Peter came up with the idea while he worked at Master Impressions and because Peter used the composite on works for Master Impressions' clients, the firm should be allowed to make the composite without having to pay a royalty. Peter, however, did not feel obligated to let Master Impressions make his composite without paying a royalty because he developed the composite on his own time and used Indianapolis Ceramics' resources. Does Master Impressions have a shop right in Peter's composite?

An application of the three factors demonstrates that Master Impressions does not have a shop right in Peter's composite, and thus will have to pay royalties to Peter to make his composite.

1. Peter did not develop the composite during work hours at Master Impressions; thus the first factor weighs in his favor.
2. The second factor also weighs in Peter's favor because Peter did not use Master Impressions' facilities to develop the composite. Instead, he used the facilities of his previous employer.
3. But because Peter introduced his composite in work he performed for Master Impressions' clients, the third factor weighs in favor of Master Impressions.

However, because the factors overall weigh in Peter's favor, and particularly because Peter did not use any of Master Impressions' facilities to develop his composite, Peter can charge Master Impressions royalties for making his composite.

6.5   Inventorship entity

The term "inventorship entity" simply refers to the individual or individuals listed as an inventor in a patent [3]. For example, if Tod is listed on the patent for Widget X as the only inventor, then Tod is the inventorship entity for the patent covering Widget X. By way of further example, if Michaela, Delaney, and Luke are listed on the patent for Widget Y as the inventors, then the inventorship entity for the patent covering Widget Y is composed of Michaela, Delaney, and Luke.

The importance of understanding inventorship entities becomes clear in the patent prosecution process. One of the grounds the USPTO uses to reject patent applications is when the subject matter of a patent application is an obvious derivation of an already existing patent [3]. However, the USPTO cannot reject a patent application as being obvious in light of an existing patent having the same inventorship entity as the patent application [3]. Some examples will further illustrate these points.

EXAMPLE 5: Ernest, Doug, and Tom are microchip development engineers for Michigan Microchip Incorporated. Ernest and Doug are listed as the inventors of Patent No. 1,100,000, which covers Chip A. Tom is listed as the inventor of Patent No. 1,200,000, which covers Chip B. Ernest, Doug, and Tom are listed as the inventors of Patent No. 1,300,000, which covers Chip C. Doug and Tom are listed as the inventors of Patent No. 1,400,000, which covers Chip D. All four patents are owned by Michigan Microchip Incorporated.

1. *Question*: If Ernest and Doug are listed as inventors of Chip E on a patent application currently pending in the USPTO, can Patent No. 1,100,000 be used as prior art to reject their patent application on obviousness grounds?

   Answer: No. For Patent No. 1,100,000, the inventive entity is comprised of both Ernest and Doug because both are listed as inventors. The patent application for Chip E also lists Ernest and Doug as inventors. Hence, because the patent application and Patent No. 1,100,000 have the same inventive entity, Patent No. 1,100,000 cannot be used to reject the patent application.

2. *Question*: If Ernest and Tom are listed as the inventors of Chip F on a patent application currently pending in the USPTO, can Patent No. 1,200,000 be used as prior art to reject their patent application on obviousness grounds?

   Answer: Yes. The inventive entity of Patent No. 1,200,000 is made up of Tom alone. The inventive entity of the patent application is comprised of both Tom and Ernest. Hence, the inventive entity of Patent No. 1,200,000 is different from the inventive entity of the patent application. Because the inventive entities are different, even though they have a common inventor, Patent No. 1,200,000 can be used as prior art to reject the patent application.

3. *Question*: Suppose that Tom resigned his position at Michigan Microchip and now works for Illinois Electronics. Assume that all of the other facts in the example remain the same. If Doug and Tom are listed as the inventors on a patent application for Chip G, which was developed as a joint project between Michigan Microchip and Illinois Electronics, and is co-owned by both companies, can Patent No. 1,400,000 be used as prior art to reject the patent application on obviousness grounds?

   Answer: No. The determination of the inventive entity is completely independent of the ownership of the patent or of the patent application. Moreover, the determination of the inventive entity is completely independent of the employment status of the inventors. Thus, because Patent No. 1,400,000 and the patent application share the same inventive entity, the patent cannot be used as prior art to reject the patent application on obviousness grounds.

6.6   Joint inventorship

A federal judge once remarked that joint inventorship is "one of the muddiest concepts in the muddy metaphysics of patent law" [4]. The problems resulting from improperly naming joint inventors on a patent application or on an issued patent can usually be avoided with adequate planning and education before any joint projects are undertaken. Hence, the selection of the individuals who are to be the joint inventors and the joint owners of an invention must be done carefully to avoid negative ramifications affecting the validity and the ownership of the patent. But before these negative ramifications are discussed, joint inventorship must be defined.

6.6.1   Definition of joint inventorship

Federal law mandates that any patent application must include all inventors, and that including more or fewer than the actual number of inventors in a patent renders a patent invalid. Joint inventorship occurs when at least two people, collaborating together, each make a meaningful contribution to the conception of a solution to a problem where the solution constitutes the invention [4]. While the definition appears simple enough on the surface, it is quite difficult to apply because some of the key terms in the definition are vague. What does it mean for two people to collaborate together — do they have to be in the same room? What is a "meaningful contribution"? How specific does a conception have to be for it to rise to the level of inventiveness?

6.6.1.1   Conception

Unfortunately there is no useful definition of exactly how or when the conception of a patentable concept occurs in the context of joint inventorship. Instead, there are two guidelines that serve to govern the boundaries of conception in a joint inventorship context:

1.   The first guideline is that one is not a joint inventor if he only suggests the desired result without providing any guidance or suggestions on how to reach this result. For example, one is not a joint inventor if he states that he would like to use a nitrogen-fueled rocket capable of taking four passengers to Mars in a voyage taking 3 months each way without providing any concrete suggestions, sketches, or plans on how to achieve this feat.

2.   The second guideline is that, for an idea to rise to the level of conception, it must include an inventive element. In other words, there must be something novel about the idea when applied within the context of the invention. An idea is not a conception if it is merely the basic exercise of ordinary skill in the art or if it is merely

the result of one following the true inventor's directions [4]. The question of whether an idea rises to the level of a conception often arises when attempting to determine whether someone, such as a lab technician or a supervisor, should be listed as an inventor on a patent application.

EXAMPLE 6: Allen is the manager for the brake design group for Vavoom Automotive. Allen had a meeting with Julie, the head of the mechanical warranty division, regarding the exceptionally high warranty repair costs for the disc brakes on their midsize vehicle line. Julie insisted that Allen and his team design and develop a more robust disc brake to be ready for production in 15 months.

At the next brake design group meeting, Allen charged his team of brake design engineers, Mark, Lionel, and Gil, to meet Julie's mandate. Lionel and Gil immediately began developing schematics for the fundamental design of the disc brake while Mark investigated the properties of different materials when exposed to high temperature, friction, and high-humidity environments. After many design iterations, Lionel and Gil designed a disc brake that combined known automotive disc technology with braking techniques used on ski lifts and with parts traditionally used to dampen noise and vibration in aircraft. No one had ever thought of this combination. When Lionel and Gil believed their design was at the proper stage to begin mock-up, they approached Mark. After reviewing their design, Mark suggested that they create the mock-up from a new combination of materials he specifically devised for this project because the materials had excellent durability and environmental tolerance.

Mark, Lionel, and Gil went to the prototype lab to speak with the lead technician, Lynn. Lynn's expertise lay in tweaking the rough designs given to her by the company's engineers. Mark, Lionel, and Gil instructed Lynn to follow the schematics prepared by Lionel and Gil and to use the material Mark developed to construct the mock-up. During the 2 weeks it took to construct the mock-up, Lynn made several technical changes to the schematics unrelated to Lionel and Gil's unique combination of known disc brake, ski lift, and noise and vibration suppression technology. Lynn was accustomed to making such changes; she had to make them on every mock-up she ever built.

After tests, results demonstrated that their design and material choice was a robust combination, Mark, Lionel, and Gil presented their results to Allen. Allen approved of the design and spent many long hours negotiating and meeting with Vavoom executives to pave the way for the design to go into production. After about 4 months of meetings and negotiations, the design was approved for production. Allen also recommended that the entire design team of himself, Julie, Mark, Lionel, Gil, and Lynn apply as joint inventors for a patent. Are Allen, Julie, Mark, Lionel, Gil, and Lynn joint inventors of the new disc brake?

No. In this example, only Mark, Lionel, and Gil can be listed as joint inventors. Lionel and Gil contributed to the invention by conceiving and designing the inventive disc brake by combining known disc brake technology with two nontraditional elements. Mark contributed to the invention by conceiving and designing a new material suited for use in a disc brake's complex environment.

Julie cannot be listed as an inventor. While Julie suggested the desired end result, i.e. a more robust disc brake, Julie did not make any intellectual contribution on how to achieve this result. She did not provide any design concepts, nor did she conceive any of the means by which to solve the problem.

Allen cannot be listed as an inventor, either. There is no doubt that Allen served an important function in the development process of the invention by spearheading the effort to push the invention through upper management. However, like Julie, Allen did not provide any insight on how to design or develop the inventive disc brake to meet Julie's mandate.

Finally, Lynn is not an inventor. Lynn's contribution is properly classified as exercising the normal skill of one in her employment position. Lynn admittedly made some changes to the schematics, but her changes were not inventive in nature. Her changes made no impact on the inventive combination of disc brake, ski lift, and noise suppression technology. Rather, the nature of the changes was more in keeping with a typical lab technician's skill set.

In deciding whether an individual is a joint inventor, it is important to not allow political pressures to color the determination. Often, the owner of a patent, such as a corporation, experiences political pressures diametrically opposed to those of the inventors. The corporate entity is pressured to name as many inventors as possible to advance its relations with its employees. Employee relations are advanced because employees perceive being named an inventor in a patent as an award of status and recognition. Conversely, because corporations often provide fixed cash awards to each inventorship entity of a patent, the inventors have a strong incentive to keep the total number of named inventors minimized. In either event, such political considerations have no place in determining inventorship status. Allowing such considerations to color the naming of the proper inventors jeopardizes the validity of the patent.

Unfortunately, many people in managerial or decision-making positions are not aware that naming too many or too few inventors on a patent renders it invalid. Rather, inventorship is perceived as being akin to authorship. It is quite common, when writing a technical paper, to list supervisors, lab technicians, professors, etc. as coauthors of the paper even if these individuals had no actual part in drafting the document. Inventorship, unlike authorship, is a legal concept. As such, naming someone as a coinventor simply to recognize his contribution to the invention is not proper;

inventorship is not an award of status or recognition and must not be treated as such.

6.6.1.2 Collaboration

To be a joint inventor, one must collaborate with the other inventors in devising the inventive subject matter [4]. Collaboration occurs between joint inventors only when they work together for a common end that is achieved by the united efforts of all of the inventors. Collaboration does not require joint inventors to work physically together. Nor does it require that the inventors work on the invention at the same time or make the same type or amount of contribution [4]. Moreover, joint inventors do not have to jointly invent every claim in the patent [4].

EXAMPLE 7: A patent has 27 claims. Of the 27, 24 claims were invented by Judy and three by Beverly. Even though Beverly's contribution to the patent is not as vast as Judy's, Beverly and Judy are considered to be joint inventors.

EXAMPLE 8: A patent has ten claims. Claims 1 to 5 were invented by both John and Doug; Claims 6 to 8 were invented by Doug alone; Claim 9 was invented by Martin; and Claim 10 was invented by Julie. John, Doug, Martin, and Julie are the joint inventors of the patent.

EXAMPLE 9: Rosalind worked in England for the International Science Foundation on the development of a microscope powerful enough to view individual strands of DNA. Rosalind and her colleague Jacques, who worked in the Paris branch for the International Science Foundation, often exchanged theories and sketches on how to design and build the microscope. Over the years, there would be periods where either Rosalind or Jacques was pulled away from their microscope development to assist with other International Science Foundation projects. Rosalind and Jacques successfully developed the microscope after nearly a decade of on-again — off-again research efforts and filed a patent application. Rosalind and Jacques are joint inventors of the microscope despite the fact that they did not work together in the same location or at the same time. Both collaborated on the same project by sharing their theories and sketches with each other to work towards a common goal. The mere fact that they were not co-located or that they did not work on the project at the same time did not affect their status as joint inventors.

6.7 Correcting inventorship errors

Sometimes, either intentionally or accidentally, patent applications are submitted to the USPTO with inventorship errors. There are two types of inventorship errors, nonjoinder and misjoinder [5]. Nonjoinder is the failure to name all of the inventors of the invention [5]. Misjoinder occurs when someone who did not contribute to the invention is listed as an inventor [5]. Neither type of error is particularly difficult to correct; rather, the ease of correction depends entirely upon how far in the patent prosecution process the error is detected.

Inventorship errors likely will not be discovered by the USPTO during the patent prosecution process because the USPTO does not have the resources to investigate and verify the inventorship of every patent application it examines. Assuming that the patent application otherwise meets all of the conditions for patentability, a patent application submitted to the USPTO with inventorship errors will likely mature into an issued patent having inventorship errors. Issued patents containing inventorship errors pose a risk of being invalidated if they become the subject of an infringement lawsuit. Obviously, to safeguard one's investment in a patent, inventorship errors should be corrected as soon as they are discovered. In most cases, inventorship errors can be corrected provided that the errors were innocent mistakes [5]. However, if an intention to deceive the USPTO regarding inventorship can be demonstrated, the patent will be invalidated, regardless of the underlying rationale for the deception [5].

EXAMPLE 10: Al, a wealthy Silicon Valley executive, invented a new widget that is expected to revolutionize nanocomputing. Knowing that an inventor is presumed to be the owner of his patent, Al decides to name his son Steve as an inventor on the patent application to avoid an increase in his tax burden. Because Al intended to deceive the USPTO about the identity of the true inventor, even though it was for reasons completely unrelated to obtaining a patent, Al's patent will be invalidated upon the discovery of Al's intentional deception.

EXAMPLE 11: Thelma is the inventor of a patent that was issued in January 2003 for Widget X. In December 2003, Thelma and Louise jointly invent Widget Y, which is an improvement on Widget X, and want to file a patent application covering Widget Y. However, Thelma and Louise realize that if they are both listed on the patent application for Widget Y, their application can be rejected as being obvious in light of the Widget X patent because Widget X and Widget Y do not have the same inventive entity. Thelma and Louise decide to name only Thelma on the patent application for Widget Y to prevent the Widget X patent from being used as a basis to reject their application. Because Thelma and Louise intended to deceive the USPTO regarding inventorship in order to obtain their patent, their patent will be invalidated upon the discovery of their deception.

EXAMPLE 12: James is a patent attorney for Eternal Cosmetics. Part of James' job is to review employee invention disclosure forms and submit patent applications for those inventions having a solid likelihood of commercial success. An invention disclosure form describing a face cream capable of eliminating acne catches his attention. James believes that such a cream would likely prove profitable, so he decides to file a patent application. While drafting the patent application, James notices that three people are listed as inventors of the face cream. Only one of the three inventors is an Eternal Cosmetics employee. The other two inventors are chemists employed by Infinity Chemical. After doing a little investigating, James realizes that Eternal Cosmetics does not have any kind of contract or agreement with Infinity Chemical addressing the ownership of jointly developed inventions. James is aware that absent an agreement to the contrary, an inventor is presumed to be the owner of his patent. Because there is no agreement between Eternal Cosmetics and Infinity Chemical defining ownership of jointly developed inventions, if James lists all three individuals as inventors, Eternal Cosmetics will have to share ownership of the patent with Infinity Chemical. James decides to name only the Eternal Cosmetics employee as the inventor of the cream to prevent any claims of ownership in the patent by Infinity Chemical. Because James intended to deceive the USPTO regarding inventorship in order to keep sole ownership of the patent for Eternal Cosmetics, the patent will be invalidated upon the discovery of James' deception.

6.7.1 Correcting inventorship errors in a patent application

Errors caught early in the patent prosecution process tend to be easier to remedy than those identified once the patent has been issued. To correct either misjoinder or nonjoinder, the misjoined or nonjoined inventor must sign a statement declaring that the inventorship error occurred without deceptive intent on his part [5]. This statement, along with other paperwork, is submitted to the USPTO, where inventorship will be corrected. However, if there is deceptive intent on the misjoined or nonjoined inventor, a correction cannot be made and no patent will be allowed to be issued [5]. Again, because the USPTO does not have the resources to investigate whether deceptive intent was present, such corrections are made as a matter of routine. Nonetheless, even if the USPTO makes a correction in inventorship that is later determined to have been based upon intentionally deceptive information supplied by the misjoined or nonjoined inventor, the patent can be invalidated.

6.7.2 Correcting inventorship errors in an issued patent

Correcting misjoinder or nonjoinder once a patent has been issued is not as simple to accomplish procedurally; however, the USPTO invariably will correct inventorship despite the patent's issued status [3]. There are two

methods by which an issued patent can be corrected once it has been
issued [3]: (1) by filing a petition to correct inventorship with the director
of the USPTO [3]; or (2) by a court order, which is described in more detail
below [3].

6.7.3 Correcting inventorship errors in an issued patent during litigation

If one is being sued for infringing a patent, he can defend the suit
by claiming that the underlying patent is invalid because of misjoinder
or nonjoinder. Successfully proving misjoinder or nonjoinder is a difficult
battle because once a patent is issued, the inventors listed on the patent
are presumed correct [3]. Thus, to overcome this presumption, a party
must be able to prove by clear and convincing evidence that the inventorship recited in the patent is incorrect [3]. The "clear and convincing
evidence" burden of proof is very high; it is similar to the "beyond a
reasonable doubt" burden of proof encountered in criminal prosecutions.
Hence, it is no easy battle to invalidate a patent based upon errors in
inventorship.

Furthermore, courts generally are quite reluctant to invalidate a patent
based upon inventorship errors on policy grounds [3]. Courts freely allow
the correction of inventorship errors as long as there is no deceptive intent
present [3]. Thus, in order for a party to invalidate a patent on the grounds
of erroneous inventorship, the party must be prepared to present virtually
irrefutable evidence demonstrating deceptive intent.

6.8 Concluding remarks

The legal status of inventor has more uncertainties than are expected by the
scientific community. Having the legal status of inventor requires a careful
application of the principles defined in this chapter to each fact-specific
situation that a scientist or engineer faces. Most scientists and engineers
who are not self-employed have agreements with their employers which
assign the employee's inventions to the employer. Because inventors are, by
default, owners of their patents, properly identifying the true inventors
impacts the patent's ownership.

The misjoinder or nonjoinder of an inventor in a patent application or in
an issued patent is correctable in most cases, but only if there were no
deceptive intent on the part of all parties, such as the inventors, management, and legal staff, involved in prosecuting the patent. If deceptive intent
is found, then the patent will be deemed invalid. Hence, to assure that a
valid patent owned by the proper entity results from a patent application,
scientists and engineers must understand how inventorship impacts both
the ownership and the validity of a patent.

References

1. Sung, L.M., Collegiality and collaboration in the age of exclusivity, *DePaul Journal of Health Care Law*, 3, 411–439, 2000.
2. Ho, C.M., Who deserves the patent pot of gold?: an inquiry into the proper inventorship of patient-based discoveries, *Houston Journal of Health Law & Policy*, 2, 107–172, 2002.
3. Chisum, D.S., *Chisum on Patents*, 89th rel., Vol. 1, chap. 2; Vol. 3, chap. 11, Matthew Bender & Co., New York, 2003.
4. Fasse, F.W., The muddy metaphysics of joint inventorship: cleaning up after the 1984 amendments to 35 U.S.C. § 116, *Harvard Journal of Law & Technology*, 5, 153–208, 1992.
5. Monheit, R., The importance of correct inventorship, *Journal of Intellectual Property Law*, 7, 191–225, 1999.

chapter seven

Internet patent document searching and interactions with an information specialist

Avery N. Goldstein
Gifford, Krass, Groh, Sprinkle, Anderson & Citkowski, P.C.

Contents

7.1   Introduction

A scientist or engineer upon hearing the word "patent" often thinks of royalties, litigation, and matters generally not of value in their work. Yet patents are a rich source of technical and strategic information owing to the statutory requirement that a patent specification teach the operative details of an invention to a level of detail understandable by one skilled in art to which the invention pertains. This requirement transcends the various patent laws between different nations and jurisdictions. An additional aspect of many patent law systems, including the U.S., is a "best mode" requirement, which mandates that a patentee recite within the specification the best mode of carrying out the invention at the time of filing. These patent law requirements combine to provide various views of the inventive subject matter, material choices, and critical relationships between components in a way that is often not found in other literature sources. The contemplated alternatives and critical parameters found in a patent often are types of information that are not found through reverse engineering of a product, even when the product is available.

While academic research is typically available in peer-reviewed journals, most corporate and sole-inventor research remains unattainable from a source other than patents. A review of the patent literature affords a perspective on the technical advances within a field, as well as strategic view as where a competitor is focusing resources and the scope of success. Patents represent a particularly valuable resource to the technical reader since an inventor is obligated to disclose sufficient technical details to enable a skilled practitioner to appreciate the details and operation of the invention. Similar detail and experimental data rarely are available from other sources.

Before the advent of computer databases, patents were not readily available to the general research community. Patent collections traditionally were stored in Patent Office archives with search options being limited. Manual searches were conducted based on a complex classification system according to subject matter, by reviewing patent abstracts, or reference to inventor or assignee names. Proximity to collections and search requirements meant that in the past only patent professionals regularly accessed patents as technical information resources.

Computer-based search systems have revolutionized patent searching, and the Internet has made the patent literature readily available to society. Improvements in database content and computer hardware have made electronic patent searching the preeminent source of patent information. As a result, manual patent searching now is reserved for specialized, exhaustive information retrieval efforts. An instance where a manual search occurs involves validity searching where references are uncovered that are relevant to one's entitlement to patent claim protection. Owing to the convenience and widespread access, a researcher or technologist can readily search electronic databases; therefore, the remainder of this chapter is spe cifically directed to electronic database searching.

7.2   Information available through a patent search

Examining the cover page of a representative patent document provides an appreciation of the information available in a typical patent database. Figure 1.1 shows a representative issued U.S. patent. The patent is reduced to a number of searchable fields, as detailed in Chapter 9, Section 4. As the vast majority of patent databases allow one to access data on the basis of field identifiers, an appreciation of field identifier content is required.

7.2.1   Patent authority

This is the authority responsible for publishing the document. In addition to patents and published applications made available by national patent offices (e.g., U.S.), there also exist regional patent jurisdictions such as the European Patent Office (EPO), and treaty organizations such as the Patent Cooperation Treaty (PCT). The patent jurisdiction typically is of concern when examining the feasibility of exploiting a technology within that jurisdiction. In the case of a PCT published application, only information about the prospect of patenting within that jurisdiction is obtained.

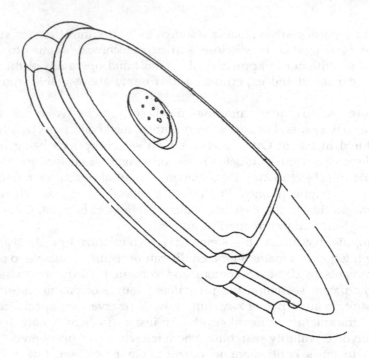

Figure 7.1 Patent drawing for a miniature flashlight.

7.2.2 *Patent/application/publication number*

This sequential number is most often a field identifier cited to obtain a particular document. This number is communicated with a two-letter prefix denoting the issuing authority. A listing of two-letter patent jurisdiction codes is provided in Table 7.1. Most notably, "WO" denotes a patent application filed within the framework of the PCT.

In some cases the patent number is followed by an alphabetical or alpha-betical–numerical code. These codes provide information regarding the document status. The code "A" or "A1" denotes a published patent application that has not yet received a substantive examination. The full set of suffix codes relating to the document procedural status is given in Table 7.2.

7.2.3 *Date of patent/publication*

This is the date on which the document became publicly available. In the case of date of patent, this date is used to determine the maximal enforce-able patent term.

7.2.4 *Title*

This field identifier defines the invention subject matter in a broad state-ment. Title field searching often is a haphazard affair, useful in locating a

Table 7.1 List of Patent Country Codes

Country Name	Country Code
Argentina	AR
Australia	AU
Austria	AT
Belgium	BE
Brazil	BR
Canada	CA
China	CN
Czech Republic	CZ
Czechoslovakia	CS
Denmark	DK
European Patent Office	EP
Finland	FI
France	FR
German Democratic Republic	DD
Germany	DE
Hungary	HU
India	IN
Ireland	IE
Israel	IL
Italy	IT
Japan	JP
Korea, Republic of	KR
Latvia	LV
Lithuania	LT
Luxembourg	LU
Mexico	MX
Netherlands	NL
New Zealand	NZ
Norway	NO
Philippines	PH
Poland	PL
Portugal	PT
Romania	RO
Russian Federation	RU
Singapore	SG
Slovakia	SK
South Africa	ZA
Spain	ES
Sweden	SE
Switzerland	CH
Taiwan, Province of China	TW
United Kingdom	GB
United States	US
USSR	SU
WIPO	WO

Table 7.2 List of Patent Kind Codes for Various Jurisdictions

European Patent Office (EPO) (EP)	A1 Publication of application with search report
	A2 Publication of application without search report
	A3 Publication of search report
	A4 Supplementary search report
	A8 Corrected title page of an EP-A Document
	A9 Complete reprint of an EP-A Document
	B1 Patent
	B2 Patent after modification
	B8 Corrected front page of an EP-B Document
	B9 Complete reprint of an EP-B Document
	TD Publication of patent claims in German language
Japan (JP)	A2 Document laid open to public inspection
	B1 Published registered patent specification without *any* A2 publication (*even afterwards*)
	B2 Published registered patent specification
	B4 Published examined patent application
	T2 Published unexamined patent application based on Internet application
	U1 Unexamined utility model
	U2 Unexamined utility model
United States (US)	A United States patent
	A1 Patent application publication within the TVPP
	A2 Patents issued after first publication within the TVPP
	A4 Patent application publication within the TVPP
	A9 Patent application corrected publication (pre-grant)
	AA Patent application publication (pre-grant)
	AB Patent application republication (pre-grant)
	B1 Reexamination certificate first reexamination
	B2 Reexamination certificate second reexamination
	B3 Reexamination certificate third reexamination
	B8 Corrected front page of a patent
	B9 Complete reprint of a patent
	BA Patent (no previous pre-grant publication)
	BB Patent (previous pre-grant publication)
	C1 Reexamination certificate (1st level)
	C2 Reexamination certificate (2nd level)
	C3 Reexamination certificate (3rd level)
	E Reissue
	E1 Reissue (pre-grant)
	F1 Corrected reissue
	H Defensive publication
	H1 Statutory invention registration (SIR)
	P Plant patent
	P1 Plant patent application publication (pre-grant)
	P2 Plant patent (no previous pre-grant publication)
	P3 Plant patent (previous pre-grant publication)

	P4 Plant patent application 2republication (pre-grant)
	P9 Plant patent application corrected publication (pre-grant)
	S1 Design patent
	X Patent document not published at all
	XB Reexamination certificate not published
	XH SIR not issued
	XT Defensive publication not published
World Intellectual Property Organization (WIPO) (WO)	A1 Publication of the Internet application with Internet search report
	A2 Publication of the Internet application without Internet search report
	A3 Subsequent publication of the Internet search report
	B1 Publication of amended claims
	B8 Second modification of the first page
	B9 Correction of a complete corrected document
	C1 Modified first page
	C2 Complete corrected document

document generic to the field. Title field identifiers are often too broad to be helpful. Additionally, title words used may be uncommon to the art or synonyms for a title keyword being searched. Title keyword searching can prove useful in identifying families of related documents.

7.2.5 Inventors

The names of individuals qualifying as inventors are provided in this field identifier. Inventor name field searches are helpful in tracking the works of an individual regardless of subject matter or assignee. This is especially true in the instance of a contractor or an individual who has changed employers.

7.2.6 Inventor address

This field identifier varies somewhat between patenting authorities with respect to the presence of street address information. Nonetheless, inventor address always includes inventor city. This field identifier is helpful in limiting inventor name searches to a particular individual when the inventor name is common to multiple individuals.

7.2.7 Assignee/applicant

This field identifier includes the name of the individual or the organization holding title to the invention. Searching this field identifier is particularly useful in determining strategic investments and technology trends within a large organization. In performing assignee/applicant searches, care should be taken to identify related affiliates, subsidiaries, and partners that may also hold portions of an organization patent portfolio under disparate names. Additionally, a change in corporate organization name associated with merger and acquisition should also be taken into account.

7.2.8   Assignee/applicant address

This field identifier provides geographic information on sites of technology ownership.

7.2.9   Application number

This number was assigned to the application upon filing and has little value as a *de novo* search parameter.

7.2.10   Filing date

This field identifier provides the date of application filing. Searching by filing date is useful when creating a temporal window of patent activity when coupled with a search of other field identifiers.

7.2.11   Priority number

This field identifier provides information on previously filed patent documents that predate and support the substance of the uncovered patent. Priority application numbers and dates afford the date of initial filing for the patent document, identify a potentially earlier date of invention, and establish a patent family genealogy. This field identifier also is extremely helpful in identifying multinational filings through a search of applications claiming the same priority numbers.

7.2.12   International patent class

This field identifier is a universal subject matter index for locating patent documents. While individual national and regional patent authorities maintain their own classification systems, there is always a cross-reference to the International Patent Classification (IPC) system. Use of the IPC system allows one to locate all patent documents relating to a specific subject matter. The IPC manual is found at www.wipo.int/eng/clssfctn/ipc.

7.2.13 National/regional patent classification

This represents another way of identifying patent documents on the basis of subject matter. National classification systems are helpful in locating older patent references that are not searchable by keywords or that predate IPC cross-reference.

7.2.14 Text fields

Keyword searches can be narrowed in most databases to particular portions of a patent document specification. Exemplary of these text fields is the invention title. Other specifically searchable text fields include invention abstract, specification, and claims. Searching keywords within specific text fields serves to narrow the number of hits to those references where the chosen keyword represents a central attribute of the technology.

7.3 Overview of patent databases available over the Internet

A researcher or technologist searching for information typically has immediate need for the desired information. The following section details databases searchable by way of the Internet, which in most instances are free to use. A cautionary note is provided that these databases all have limitations in searchable content, image retrieval, and available Boolean operators. The net result is that a given search should not be considered as authoritative, even though considerable information can still be found.

An overview of several patent databases follows. One should not expect to uncover the entire state of the art by searching in a given national patent database. In reviewing the following databases, the reader is cautioned to identify those portions of a patent document that are in fact searchable. For example, while the United States Patent and Trademark Office (USPTO) database allows one to search the full text of patent documents, other databases provide only for title keyword searching. Keyword searching in general is limited only to those documents that have been converted to a searchable format. Often, patent documents will be available through patent number or classification searches that will not be identified through a keyword search.

Patent searches are readily performed through a search link. The link and an overview of the web-based search capabilities for each site follows.

7.3.1 European Patent Office (ep.espacenet.com)

The EPO contains bibiliographic data for all patents published in any European Patent Organization member by the EPO and World International Property Organization (WIPO) currently dating back to 1998. Access is also provided to a comprehensive collection of patents published worldwide since 1920. For patents published since 1970, each patent family in the collection has a representative document with a searchable English-language title and abstract. A user can display and print out bibliographic data including the abstract, and where available, the cover page image together with any drawings and full text in machine-readable form. This is probably the most comprehensive free patent search site available on the Internet.

7.3.2   Japan (www19.ipdl.jpo.go.jp/pa1/cgi-bin/pa1init)

Patent abstracts of Japan are searchable from 1976 onward based on a variety of search fields such as keywords, international classification number, publication number, patent number, and appeal number. Those abstracts published after 1992 include legal status information as well. The Japanese patent system has employed several numbering systems for patents and publications over the years that are not necessarily intuitive, and care should be taken that searches based on these criteria are in fact accurate. Patent abstracts of Japan are written in a format including the problem to be solved and the constitution of the invention. This format affords additional modes of performing successful keyword searches. If a searcher has a language preference other than Japanese, with the discovery of a relevant Japanese abstract, a family search can be undertaken on one of the other websites detailed in this chapter to determine if there is a corresponding application or patent in a preferred language. The Deutsches Patent Informations System (DEPATIS) is an exemplary and free patent family search tool. A recently added tool provides on-the-fly translation from Japanese to English of whole patents.

7.3.3   United States Patent and Trademark Office (www.uspto.gov/ patft/index.html)

The USPTO website includes U.S. patents from 1790 forward and published applications from 2001 forward. Patents are searchable by fields from 1976 forward. Published applications are searchable by fields from 2001 forward.

7.3.4   Germany (www.depatisnet.dpma.de)

DEPATISnet allows an online search of a variety of ever-expanding international patent publications compiled in the DEPATIS database. This database is the internal patent information system of the German Patent and Trademark Office. In addition to the collection of German patent resources, the information compiled in DEPATIS is in the original language. For example, Japanese patent abstracts are imported in English, French patent titles and abstracts in French. As a result of the multiple language content of DEPATIS, search terms must be designated in the appropriate language or multiple language searches conducted. A searcher can use a patent family search mode to find related patent documents within a particular patent authority and in different nations. The patent family mode is particularly helpful in finding corresponding documents in a language other than that uncovered.

7.3.5 Canada (patents1.ic.gc.ca/intro-e.html)

The Canadian Intellectual Property Office offers access to more than one-and-a-half million Canadian patent documents spanning more than 75 years. English and French language searches are available and represent a good source for determining French–English keyword pairs.

7.3.6 Russia (www.fips.ru/ensite)

In spite of the wealth of Russian language technical information that has been published, on-line searches for technology developed in the former Soviet Union remain difficult. The Russian Patent Office provides access to Russian patent abstracts and drawings dating to 1994.

7.3.7 World Intellectual Property Organization (http://ipdl.wipo.int/)

A searcher has access through this database to full text PCT documents from 1998. A full field text search tool is available in the three official WIPO languages: English, French, and Spanish. As PCT applications are filed on technology where the applicant is potentially going to seek international protection for the invention detailed in a PCT application, this is an excellent source of information about competitive organization patenting strategies when compared with national patents granted based on the PCT filing.

7.3.8 Commercial databases

There are a number of subscription accessible databases providing a searcher with a compendium of information that otherwise requires searches across multiple patent authority websites. In addition, these sites typically offer enhanced search fields, period search alerts, and access to related information such as litigation history, maintenance fee status, and updated assignment information. The fees structures and relative merits of a particular commercial database are unique to each searcher. Without endorsement of any of these databases some well-established information providers include Delphion, Dialog, and Lexis.

7.4   Performing various types of searches

Patent searches are performed most often for either informational or legal purposes. An informational search is intended to derive specific information contained within patent documents. The basic procedures for performing various informational searches will be detailed in this section. The reader is assumed to have a working knowledge in the use of Boolean search operators such as AND, OR, NOT and proximity operators in order to write a search query. Before conducting a search on a given database, a user is encouraged to review the database help section in order to get familiar with the available operators and the associated retrieval parameters.

Patent searching involves looking for information that may or may not be present in a given database. The extensiveness of a given search depends on the use of the resulting information, as well as the resources available. While an informational search is concerned with patent document content, legal searches focus on the scope of protection afforded. A legal search frequently is accompanied by a patentability or infringement opinion generated by a patent attorney.

The sections that follow summarize the characteristics of various types of searches. Where practicable, a search protocol is provided for a given search type.

7.4.1   Informational searches

7.4.1.1   Identifier searches

The information fields detailed above represent valuable information that in the era of manual searching required specific methods to investigate. These are now problems of the past. Using the advanced search fields of virtually all databases, identifier field searches to find, for example, assignee information, or the address of an inventor, is merely a matter of entering the known information into the appropriate field and reviewing the hits for the desired information.

7.4.1.2   State-of-the-art searches

The purpose of a state-of-the-art search is to determine what has previously been done by others to solve a particular problem. This type of search is frequently attempted after identifying an area of particular interest, yet before devoting significant resources to the endeavor. With a state-of-the-art search completed, the various solutions to a particular problem can be evaluated, as well as participants in the field. Care should be taken in performing such a search to adequately define the area of interest commensurate with the desired information. While a broad concept is likely to retrieve hundreds, if not thousands, of relevant patent records, a narrow definition may well miss relevant references within the area. As such, a layered search strategy iterating between keywords and/or classification is most often recommended to strike a proper balance.

To retrieve patent records relating to a specific area, it is important initially to create combinations of keywords, classifications, and/or bibliographic information such as assignee or inventor name. An optimal search strategy involves combining criteria matching characteristic concepts and iterating the criteria in the course of the search.

7.4.1.2.1 Step 1. A state-of-the-art search begins with listing search criteria, which include:

A. What does the invention do?
B. How does the invention work?
C. What elements are required?
D. What is the inventive result including drawbacks?
E. What is the method by which the invention is produced? (if applicable)
F. Step 2

7.4.1.2.2 Step 2. The next step is to translate the answers to the above questions into keyword search statements that take into account synonyms, misspellings, truncations, and possible grammatical forms in which a keyword might be used. At this stage in the search it is better to be over inclusive by choosing general keywords to define a concept. If the database being used so permits, it is recommended to order search results in a format other than the default of the first chronological order. For instance, grouping initial search results by classification identifies subject matter commonality between references. Alternatively, grouping keyword search results by assignee or inventor offers a coherent progression of research efforts by the entity. The limitations of the database need to be considered in evaluating the search results. To date, none of the databases detailed herein is able to access all patents within the collection by way of a keyword search. If the technology being searched predates the records available with the database, a patent-searching professional will likely need to be consulted.

After parsing search results from an initial keyword search, often one can eliminate a large number of references based upon examination of patent abstracts and titles. The remaining records recovered can be labeled as relevant or questionable based upon this limited text review. If an especially relevant document appears, it is worthwhile to identify the IPC and/or national patent classification codes. Examination of a highly relevant classification code definition in a classification manual and related classes often serves to refine the search, especially upon combination with keywords. Alternatively, one can simply retrieve all the documents within a specific classification.

7.4.1.2.3   Step 3.   A classification search is conducted based only upon classification codes for the invention described in the first step. Such a search begins with an examination of a patent classification index. The index broadly identifies the class in which the invention might be found.

7.4.1.2.4   Step 4.   Examination of the manual of classification for the relevant database to identify broad classes and specific subclasses relevant to the invention. Care should be taken to recognize that any given record often has cross-referencing classifications as many inventions relate to different technology areas. For example, for a pharmaceutical, the process of making the active ingredient and the process of formulating the pharmaceutical composition will each have a separate classification. The IPC manual is found at www.wipo.int/eng/clssfctn/ipc. The United States Manual of Patent is found at www.uspto.gov/web/patents/classification. A concordance between these two classification systems is found at the latter link.

The classification search results are then reviewed and binned according to irrelevant, questionable, and relevant, based upon examination of patent abstracts and titles. Once again, for particularly relevant records, related classification codes are noted as are keywords.

7.4.1.2.5   Step 5.   Search reiteration continues with refinement of classification and keywords, and combinations thereof until a closed universe of relevant records is identified. As continued searching shows only previously examined references and new references that are less relevant than those already identified, it is an indication that the search is approaching a point of diminishing returns.

7.4.1.2.6   Step 6.   Upon completing the combined keyword and classification search, it is recommended to perform a refined pure keyword search to detect the rare document with an incorrect or missing classification code.

7.4.1.2.7   Step 7.   The full text or full image versions of the most relevant documents are now reviewed in detail and possibly used to further refine the search strategy.

The maintenance of a log detailing search strings and results is recommended if the search is ever to be updated or upon referral of the search for additional work to an information specialist.

7.4.1.3   Novelty search

A novelty search is designed to locate the references that anticipate an invention or to determine if a patent filing is justified based on the likely claim protection available. The scope of the search includes all information in printed publications or in public use before the invention/filing date. The distinction as to whether a particular reference constitutes prior art is a mix of factual and legal determinations best left to a patent agent or attorney.

The importance of an invention date versus a filing date is a matter dictated by the patent laws of a jurisdiction. Most countries consider the filing date as the critical date for determining prior art. The U.S. is in the minority in treating the date of invention as the seminal prior art cut-off date, and in recognition of this emphasis on invention date affords up to 1 year before the filing date of a patent application as the date of invention for purposes of determining whether a given reference is in fact prior art.

A novelty search that fails to uncover a reference that encompasses an inventive concept also serves to identify references that, when combined, may render an invention an obvious extension from the information detailed in the prior art as a whole. This is known in the terminology of various patent jurisdictions as inventive obviousness or lacking an inventive step. This notion is the legal aspect of a novelty search and is best discussed in the context of a specific example with a patent professional.

It is important to note that a novelty search will not provide information regarding the ability to practice the invention. By way of example, if one were to develop an invention that involved combining components A + B + C, a novelty search would uncover references including A + B, A + C, A + B + C. An enforceable patent relating to B would be infringed by the new invention A + B + C, but would not be identified by the novelty search.

Typically, an information specialist is charged with conducting a novelty search by examining patent databases. A search of the patent literature is necessarily incomplete, but is an area frequently not reviewed by technologists. Patent databases are often the source of the most extensive prior art teachings since ranges and constructions not actually used are detailed in the broadest operative terms within a patent document. Before the advent of computer database searching, it was not common that multinational collections were searched as part of a novelty search. Now it is a feature of a quality novelty search that at least European, U.S., WIPO, and Japanese patent documents be searched. Owing to the concentration of certain technology areas and industries in particular countries, other specific national patent collections should be examined in performing a proper novelty search.

Inherent in a thorough novelty search is the presumption that the technologist will provide the most relevant open literarture references, since the cost of conducting a search of the open literature is often prohibitive at this juncture in the patent process. The technologist may also assist the novelty search by providing the names of known workers and organizations working in the inventive field.

Attention must also be paid to the nature of the invention as to which databases should be accessed for a novelty search. For instance, chemical inventions, especially those including structures, are often most readily uncovered through examination of a technology-specific commercial database such as the American Chemical Society (CAS). Electronic business methods as a new area of patentable technology are most likely to be found through an open Internet search. In instances where prior art is most likely found in a nonpatent database, the patent literature is searched for the added breadth and operative technical teachings that are characteristic of the patents. Gene euclidation or medical treatment–based inventions merit a search in the National Library of Medicine (http://www.nlm.nih.gov).

7.4.1.3.1   *Step 1.*   A novelty search begins with a listing of potentially patentable aspects of the invention. Potentially novel features should be listed in broad terms with optional element or preferred forms being delinieated as subinventions. Such a list not only facilitates searching the various inventive aspects but also helps a patent agent or attorney in subsequent claim drafting to define the scope of the patent monopoly sought. The listing should be divided into novelty categories as applicable of:

- material (device/compound/article);
- process of production, whether of a known or novel material; and
- process of use.

It should be apparent at this point that a single invention actually may exhibit several independent bases of novelty and therefore may constitute multiple inventions for the purposes of a novelty search. As a patent reference is placed into a classification system based on the subjct matter of the claims, a patent only claiming a material is unlikely to be found in a search-based process classification. In recognition of this ambiguity, patent classification systems include cross-references to other classifications. However, owing to a subjective component of applying cross-references and indeed primary classification, sole reliance on classification often is an inadequate search strategy.

7.4.1.3.2   *Step 2.*   A keyword search is conducted with the knowledge that keyword searches are limited only to recent portions of patent collections and can miss even recent references that use synonyms for the selected keywords. The goal of the initial keyword search is to provide some initial guidance in selecting classifications. The classification and cross-classification of the most pertinent references are noted. These serve as a starting point for searching manuals of classification to determine where the closest references may reside.

7.4.1.3.3   *Step 3.*   Examination of a manual of classification is then performed for the relevant database to identify broad classes and specific subclasses relevant to the invention. The object of this step is to find a handful of productive subclasses since a given class might include tens of thousands of individual patents, and a subclass often contains tens to hundreds of patents. Upon identifying a few promising classes, a tedious search through all the references within that subclass is conducted with particular note being taken of highly relevant references for subject matter and classification information. Often a considerable time is saved in searching at least U.S. references through the use of simultaneous classification codes assigned from U.S. and International Classification systems.

7.4.1.3.4   *Step 4.*   Relevant subclasses are searched in their entirety for relevant references. It is important to note that this is a methodical and slow process. A search can also be performed with some timesaving by reviewing references in classification order as opposed to chronology. Even though a number of references are contained within a certain subclass, they have not all arrived there in the same way. For instance, as a classification system evolves, certain subclasses are merged, eliminated, or added and the reference classification is changed. These vestigial classification codes are often helpful in eliminating sizeable portions of a relevant subclass. This organization of references means that if a patent publication title and a given subclass are unproductive search areas, then like classified references can be bypassed.

In the case of U.S. patent reference, searches resorting to IPC codes should also be considered, as it is possible that the specifics of the international classification system may group references in a way that is more akin to the invention being searched than does the U.S. classification system.

Additionally, searching for documents having simultaneous placement in two or more subclasses is an effective way to narrow the documents subject to review. However, care should be taken that cross-referencing is practical only as an initial search strategy or only after a level of confidence has developed that potentially meaningful documents are not being overlooked.

As before, information about search hits is noted. After exploring a number of promising classes/subclasses and discarding those that proved unproductive, those references that appeared most promising are revisited.

7.4.1.3.5   *Step 5.*   A patent includes citations to those patent and non-patent publications uncovered by the granting patent office or provided by the applicant. By following up with a search of the references cited in a relevant patent, still other relevant references are uncovered. These cited references in turn can be traced to review the references cited therein. Most free websites do not have an automated method of performing a cited reference search, meaning that cited references will have to be manually entered into the search interface as patent numbers. A notable exception is the USPTO link that provides hyperlinks to at least cited U.S. patent references.

7.4.1.3.6   *Step 6.*   The process in Step 3 to Step 5 is repeated in a feedback loop using even better references until a closed universe of references becomes apparent for an inventive concept as delineated in Step 1. In other words, searching by keyword, class/subclass, and reference citation leads to the same group of references. This is a strong indication that the most relevant references have been uncovered. However, since the search is directed to an invention that is believed to be novel, there is a real possibility that there are no relevant references.

7.4.1.3.7 Step 7.   Record the complete search history with an indication of the database, search strings, class/subclass information. This information is helpful in subsequent refinement of the search.

7.4.1.3.8 Additional comments on novelty searching.   The results of a novelty search are clear in the extrema where either nothing of relevance is found or a single document wholly discloses the invention. In actuality the vast majority of searches yield results indicating that there are documents that disclose relevant portions of the searched invention, but they are incomplete in fully teaching an invention or the invention in combination with one or more subinventions. This means that the inventive subject matter has passed the patentability threshold of novelty. Novelty is an objective standard and one where a technologist familiar with the invention subject matter is well suited to make such a determination.

Obviousness, or the inventive step component, in patentability is a subjective standard for patentability that is difficult to assess without the inputs of a patent professional. The basis of the obviousness patentability standard is that while an invention may not have been made before the deviation from that done previously, it is a simple extension not worthy of protection. For example, the replacement of an incandescent bulb with a light-emitting diode in a flashlight, or a nail with a screw for the hanging of a picture, if not known at the time of replacement, are different but lack creativity in both form and function. While the obviousness of an invention over a single reference is fairly straightforward to apply, the standard is a comparison of the invention relative to the total knowledge of the field as of the critical date, and as such multiple references can be combined. The legalities of what constitutes a proper refererence combination and the level of the teaching found in a document are a matter that requires the input of a patent professional. Differences between national patent laws further complicate this assessment.

With a patent search indicating invention novelty, but possible issue relating to the obviousness of the invention, the decision as to whether the invention constitutes subject matter for a patent filing becomes a multifaceted decision-making process that ultimately is a business decision with inputs from the patent professional, and marketing and technical staffs. Other avenues of protection include maintaining the invention as a trade secret, statutory invention registration, or trademark protection. A detailed discussion of some alternate forms of intellectual property protection are found in: *Intellectual Property: The Law of Copyrights, Patents and Trademarks* by Roger E. Schechter, John R. Thomas, Jay Thomas (Thomson West, 2003).

Because a novelty search involves a quest for materials that might not exist, not finding a novelty-destroying reference can either be attributed to an incomplete search or to the fact that the document simply does not exist. In order to gain confidence that a novelty-destroying reference does not exist, it is important to record the databases and search history used in the context of a given database. As all of the free Internet-based patent authority databases and subscription-based commercial databases have gaps in the searchable collections and the search strategies available, the nature of the search a technologist has performed greatly enhances the efficiency of the search performed by an information specialist.

7.4.2   Legal searches

While a novelty search has legal aspects, there are other types of searches whose function is overwhelmingly legal in nature. Legal searches are most often conducted after a patent has been granted, and your interest is either in bolstering or destroying the monopoly granted under the patent. As the stakes of such a search are much higher than searches conducted prior to patent application, a legal search is of broader scope. In short, a legal search requires database resources and often manual searching across several languages and document collections that are beyond those available to a technologist. The following discussion is not intended as a guide to performing such a search, but rather to familiarize the reader with the basics of such searches so that he or she is equiped when called upon to assist an information specialist.

7.4.2.1   Validity search

A validity search is conducted to ascertain if references relevant to a patent grant were not considered in the course of patent prosecution. For a patent to be granted it must be novel, and therefore a document that existed prior to the critical date for a patent that was not considered in the granting of the patent calls into question the validity of the patent.

The circumstances when a validity search are conducted are monumental in the life of a patent. The high cost of a validity search becomes justified in those instances when the value of the monopoly conferred by the patent is questioned through licensing or infringement litigation. The validity search represents the most extensive and costly of all patent document searches. It is justified in the following circumstances for different participants in the patent arena.

A potential licensee or purhcaser should conduct a validity search to determine if the monies about to be paid to be able to practice a patented technology are being well spent. Even if a validity search does not preclude a licensing arrangement, a potential validity issue can be used by a licensee to negotiate more favorable licensing terms. A validity search protects a licensee from a third party being able to operate at a competitive advantage relative to the licensee by doing business without the burden of having to pay a royalty on a technology that is in the public domain or not under the the monopoly of the licensor.

Similarly, a patent holder may conduct a validity search prior to licensing or attempting to enforce a patent against a potential infringer in order to assess the strength of patent claims. If a patent holder later uncovers documents that indiciate that the patent scope is not contiguous with the prior art, there are mechanisms available in the various patent jurisdictions to expand or narrow the claim scope. In the U.S., reissue and reexamination procedures exist within the PTO to correct claim scope in light of newly discovered documents.

A market competitor who has been charged, or fears being charged, with patent infringement has an enormous financial motivation to render the patent in question invalid and therefore remove the potential liability. Based on the liabilities involved, validity searches costing hundreds of thousands of dollars have been commissioned and a legion of information specialists are scouring libraries around the world. The value of a successful invalidating search is such that there are bounties posted for effective invalidating documents (www.bountyquest.com).

A validity search represents an extreme version of a novelty search that must be comprehensive. The goal is to find at least one anticipatory reference that encompasses the patent monopoly defined by a patent claim. As such a validity search looks for references that predate the critical date of the patent. The search is not limited to patent documents or language. The fact that the patentee had no knowledge of the reference and independently reinvented the technology is of no consequence. Knowledge of the prior art is not important because the patent system is based on the premise that the creator of new technology should be rewarded with a limited-term monopoly in exchange for sharing the practice of the invention with society. The modern patent system will not take something from the public and give it to an individual by way of a patent.

The discovery of potentially invalidating documents is quite common owing to the limitations of prior art searching by both inventors and patent authorities in the era before computer-accessible databases. Even with the advent of database searching it was all too common that keyword searches were only performed that limited the documents searched to those with keyword-searchable fields. Additionally, access to foreign language and multiple patent authority databases left many areas unsearched. Nonetheless, a validity search is the proverbial search for a needle in a haystack with little to guide an information professional to find the needle at the top rather than at the bottom of the haystack.

The technologist can serve an important role in the validity search. The private literature collection a technologist consults for procedures, known results in the field, and keeping track of competitive groups in the field represents an excellent place to uncover that one reference that may well impact the validity of a patent in question. Additionally, a technologist may well have copies of old catalogs, technical bulletins, and trade publications that frequently fail to enter searchable databases. It may even be the case that a patentee may publish a technical bulletin or provide materials in the course of a sales presentation that negatively impacts the patentee's ability to obtain a valid patent. For those in expectation of a validity search being performed regarding a subject familiar to a technologist the following steps be undertaken:

- Collect and archive secondary technical materials such as catalogs, technical bulletins, and meeting abstracts that contain information detailing the workings of relevant products; these materials should be kept for a term of at least 22 years for validity purposes and longer for novelty purposes.
- Retain contact information of sales people, collaborators, and colleagues who may also have technical knowledge that will be relevant to a subsequent validity search.
- Discuss the object of a validity search when commenced with an information specialist to determine if the desired subject matter is familiar or even found within the secondary technical materials being collected. With the guidance of patent counsel, use the contact information to expand the number of human sources who may have knowledge of references relevant to the validity search.

The interaction between a technologist working in the area of a validity search and an information specialist conducting the search is often the quickest route to a successful search. Secondary references that are unlikely to be included in electronically searchable databases represent a source of validity impacting information that cannot be discounted. The importance of these secondary references to a validity search underscores the value of making these references available during a novelty search and of record at the time of filing of a patent application to preclude the drama associated with reviewing these references in the course of a validity search.

7.4.2.2 Infringement search

An infringement search is conducted to examine the claims of enforceable term claims of patents in the relevant jurisdictions. The purpose of an infringement search is to prospectively avoid a charge of patent infringement through an assessment of whether a product or process to be produced falls within the language of an enforceable term patent claim. Various forms of an infringement search that depend somewhat on the object and the scope are known as "clearance" and "freedom to practice" searches. The details of these variations are immaterial to the overview of legal searches that this text is intended to provide.

The scope of an infringement search is considerably different than the other types of searches. An infringement search examines the claims and not the complete teaching of the patent. Since infringement is a concern where the product or process will be made, used, or sold, only those countries where one plans activity need be searched. A patent that has expired through running the statutory life or failure to pay maintenance fees is disregarded as not being a basis for a charge of patent infringement. For this reason, the scope of an infringement search is only the past 20 years. An exception to the 20-year search range of drug patents in some jurisdictions such as the U.S. where the patent term is tolled for a period of time while the patent subject matter undergoes regulatory approval.

An infringement search is an expensive and time-consuming endeavor that requires both an authoritative search of the active term portion of a national patent collection and a legal review of the claims identified. The justification for this effort is that the cost of discovering a potential infringement problem initially is far less than the cost of contending with litigation after commercialization. Potential investors, to assure that developing technology can be successfully commercialized and therefore a return on investment is possible, also conduct this type of search.

A technologist has a role in defining the parameters of an infringement search. This is actually the stage where most infringement searches fail to perform the desired goal of litigation avoidance. A search that is crafted too narrowly and directed to the product or process as a whole will miss enforceable term claims directed only to an element, ingredient, or process step used in the product or process being investigated. It is here that a technologist offers essential insights into the background of particular components and their sources. The vendor of a component used in the product or process of interest can be looked to to assure that the component is delivered free of patent infringement. Otherwise, it is common to ask a vendor to indemnify the purchaser against charges of patent infringement associated with the use of a bought component for a specified purpose.

Component pieces produced according to internal specifications and the processes by which those components are made represent sources of potential infringement. For example, in building a welded door assembly, the assembly as a whole, the welding techniques, the various component pieces, and the sequence of assembly steps all need to be cleared as part of a satisfactory infringement search. While an information specialist can perform a painstaking search on each of these aspects without any prior knowledge of the field, a technologist can greatly streamline the search by identifying aspects of the product or process of interest that are so old as to not possibly be the subject of an enforceable term patent claim. A technologist through a meeting with an information specialist or patent attorney can greatly reduce the complexity of an infringement search by identifying old aspects. The technologist invariably is asked to provide the patent attorney with supporting documentation before the attorney is able to render an opinion as to the old aspects. The reason an old aspect precludes infringement relates to validity: if practicing old art is alleged to infringe a patent claim, then the patent claim is invalid as encompassing prior art. For this reason, the practice of technology detailed in an expired patent is a particularly safe harbor from a charge of patent infringement.

Once a patent claim is potentially infringed the claim scope must be interpreted. The patent attorney relies on the laws and interpretative tools available in a particular jurisdiction. In the U.S., claim interpretation involves an examination of the patent prosecution history that includes all official correspondence between the PTO and the patentee. If the proposed product or process cannot be distinguished from the claim after interpreting the claim scope, then one has the choice of conducting a validity search to determine if the offending claim is likely to be adjudged valid, redesigning the product or process to avoid the claim coverage, or seeking a license to practice the claim.

7.5 Tips on preparing search inputs

7.5.1 Keywords

Most Internet databases limit one to the Boolean operators AND, OR, and NOT. The NOT operator should be used with great care since a reference in an invention background may well contrast to prior systems that include the exclusionary term. These operators are far less sophisticated than proximity operators that allow for terms to be searched within a specified number of words, sentences, or paragraphs. Proximity operators are found on examiner search systems. There are a limited number of locations where individuals are granted access to these search systems. Each patent jurisdiction website provides information about remote site access to these search systems. In the U.S., access to Examiners' Automated Search Tool (EAST) can be had through facilities located in Detroit, Michigan, and Santa Clara, California.

The choice of keywords is subjective. As different patent drafters will chose different terms for the same object, one should not expect to find all the relevant art through a keyword search. Rather, the goal is to find a relevant set of possible classifications by way of a keyword search. The greatest asset in constructing keyword search phrases is a thesaurus. Exploring synonyms for an inventive element gives one the greatest chance of finding a relevant reference and therefore relevant classification information. Searches can be greatly narrowed to high probability hits by looking for a keyword in a particular field such as an invention title, claims, or abstract.

Table 7.3 The European Classification System Extended to Completion for a Representative Example

Class	Description
A	**Human necessities**
A01	
A01B	**Soil working in agriculture or forestry; parts, details, or accessories of agricultural machines or implements, in general** (making or covering furrows or holes for sowing, planting, or manuring A01C5/00; soil working for engineering purposes E01, E02, E21 [N: measuring areas for agricultural purposes G01B])
A01C	**Planting; sowing; fertilising** (combined with general working of soil A01B49/04; parts, details, or accessories of agricultural machines or implements, in general A01B51/00 to A01B75/00 [N: apparatus for spreading sand or salt E01C; sowing and fertilising with aircraft B64D1/16 to B64D1/20])
A01C1	**Apparatus, or methods of use thereof, for testing or treating seed, roots, or the like, prior to sowing or planting** (chemicals for A01N25/00 to A01N65/00 [N: irradiation in general B01J19/08])
A01C1/00	As above
A01C1/00B [N: potato seed cutters]	
A01C1/02	Germinating apparatus; determining germination capacity of seeds or the like (germinating in preparation of malt C12C [N: C12C1/04]) [C9508]
A01C3	**Treating manure; manuring** ([N: ploughs with additional arrangements for putting manure under the soil A01B17/00B]; dung forks A01D9/00; organic fertilisers from waste or refuse C05F)
A01C5	**Making or covering furrows or holes for sowing, planting, or manuring**
A01D	**Harvesting; mowing** [N: (parts, details, or accessories of agricultural machines or implements in general A01B51/00 to A01B75/00)]
A01F	**Processing of harvested produce; hay or straw presses; devices for storing agricultural or horticultural produce** (devices for topping or skinning onions or flower bulbs A23N15/08)
A21	**Baking; edible doughs**
A22	**Butchering; meat treatment; processing poultry or fish**
B	**Performing operations; transporting**
C	**Chemistry; metallurgy**
D	**Textiles; paper**
E	**Fixed constructions**
F	**Mechanical engineering; lighting; heating; weapons; blasting engines or pumps**
G	**Physics**
H	**Electricity**

7.5.2   Determining relevant classification

7.5.2.1   International patent classification

IPCs are common to nearly all patent systems. IPCs can be accessed by looking for conceptually descriptive keywords. In this instance the use of the WIPO electronic version of the keyword index is preferred since the classification symbols listed in the index are hyperlinked to full definitions and structures where relevant. The European patent classification (ECLA) builds on the IPCs to create an extensive and precise derivative indexing system.

IPC symbols are organized in a hierarchy of eight sections that are delineated into classes, subclasses, and finally subgroups. A more detailed derivative system is the ECLA. An example of the ECLA system is shown in Table 7.3.

7.5.2.2   United States Patent classification

The U.S. patent classification system also has an index delineated alphabetically according to keywords. This is an excellent place to begin a generic or broad search, with the knowledge that only U.S. patents and published applications will be so retrieved. The index to the U.S. patent classification system references the appropriate class and subclass, which are delineated into telescoping subclasses. Use of the manual of classification must also include a check of classification definitions including a review of the "statement of class subject matter." Further review of the notes relating to particular subclasses, the original classification, and current classification are also checked since the U.S. patent classification system is quite dynamic with periodic reorganization and reclassification, as particular subclasses become unwieldy and the definitions technologically obsolete.

7.6   Exercises

- Find the U.S. classification and IPC for a lint roller. Compare your finding with those found on the cover page of U.S. 6,055,695 reprinted in Figure 1.1.
- Find the first patent issued to Robert Jarvik for an artificial heart. Have equivalents of this patent been filed in other countries?
- In what country was the earliest filing made for bamboo charcoal soap?
- Find the family of patents for anatase containing white paint that has a priority date in the early 1990s.
- Find a patent for a miniature flashlight that contains Figure 7.1.

chapter eight

Interactions with a patent agent or attorney

Tom Brody
Coudert Brothers LLP

Contents

8.1   Introduction

A patent provides the patent owner with a right to exclude others from making, using, or selling a claimed invention. "Claimed invention" means an invention as described in the claims. Ownership of a patent does not provide ownership of any invention, device, or apparatus. All that is owned is the patent, where this ownership provides the owner with the above-mentioned right to exclude. In addition, patents can help persuade investors to fund the refinement, testing, and commercialization of an invention. As an early step in acquiring a patent, the inventor needs to supply various types of information to the attorney or agent. In turn, the attorney or agent writes the patent application and files (submits; mails) it with the United States Patent and Trademark Office (USPTO). At this point, the work is only halfway complete. The examiner reviews the application, and mails a document called an Office Action to the attorney. The Office Action usually consists of a few pages of critical comments, called rejections. The rejections are based on various statutes, such as 35 U.S.C. 102 (a "102-rejection") or 35 U.S.C. 103 (a "103-rejection"). It is the responsibility of the attorney or agent to respond to the rejections, preferably with input or feedback from the inventor. If the attorney neglects to rebut these comments, or writes an unconvincing rebuttal, the rejections stand and the patent is not allowed.

The attorney or patent agent reads published court cases and from them acquires guidance in writing patent applications and in formulating rebuttals to the rejections. These publications document court decisions from the U.S. Supreme Court, the Federal Circuit, and the U.S. District Courts. The most useful sources of guidance are the decisions from the Federal Circuit (Fed. Cir.). An outline of how the Fed. Cir. makes its decisions is available [1].

This chapter applies to U.S. patent law, and should not be used as a guide when the goal is to apply for and acquire a patent in any foreign country, e.g., England, France, Germany, Japan, or Australia.

8.2   Consultation

The inventor needs to provide the attorney or agent with a description of the invention, a description of the working environment of the invention, and copies of the prior art. The prior art includes patents, publications, newspaper articles, and advertisements. The attorney's job is to compare the invention with the prior art, determine the novel features of the invention, and write claims in a way that assures maximal protection of these novel features. Writing the claims is often like a balancing act. In writing claims of optimal scope (claims of optimal coverage), the attorney covers at least one of the novel features of the invention in each independent claim, where the claims are written in a general style, in order that later-arising technologies can fit into the description. At the same time, the attorney attempts to avoid writing claims that describe devices or methods that are already patented, published, or that are generally known.

In addition to providing the attorney or agent with a description of the invention and its novel features, and the prior art references, it is useful for the inventor to provide the attorney with information regarding the inventor's activities in selling, public use, public experimentation, or publication.

Beyond the initial consultation, the inventor may provide input to the attorney regarding Office Actions. The inventor may aid the attorney in the formulation of responses to the Office Actions. For example, the attorney may require advice from the inventor on which elements of the invention are the most valuable to the inventor or to his or her company. Or the inventor may provide advice as to whether it is acceptable that certain elements of the claim be changed, where the result is narrower claim coverage. In responding to Office Actions, it is the responsibility of the attorney or agent to respond in a way that maximizes or preserves the coverage of the claims, while avoiding making amendments that narrow the coverage of the claims. However, it is sometimes the case that some or all of the features described in the claims already exist in the prior art, and the attorney may have no choice but to narrow the claim coverage or to acquiesce in the examiner's rejection.

8.2.1   *Regarding activities*

Before delving into details of the invention and its innovative features, the best way to begin the consultation is to determine if the inventor has sold the invention to any party, if the inventor has used or shown the invention in public, or if the invention has been described in any publication. If the inventor has already sold the invention, has already used or shown the invention in public, or has published a description of the invention, the attorney must then determine when the use, sale, or publication first occurred. If the date of first sale, use, or publication occurred prior to 1 year before the expected filing date of the patent application, then the invention cannot be patented [2, 3]. In other words, where the inventor has sold the invention, publicly used the invention, or published a description of the invention prior to 1 year before the date of filing the patent application, the attorney should refrain from preparing a patent application. Attention to the filing date is a major issue in patent law.

Another activity of concern is whether the inventor has documented the date of conception of the invention, and whether the inventor has documented the ensuing activities in developing the concept to a working model or prototype. Documenting the date of conception in a laboratory notebook, and the subsequent more-or-less continual work on the invention is usually not required to obtain a patent. However, where the examiner determines that another patent application exists (where this patent application claims the same invention), the inventor's laboratory notebook records may be critical to the survival of the patent application. The date of conception is a major theme in patent law.

A further activity of interest is any commercial success of the invention. The inventor should keep note of any commercial successes, as these may prove useful in formulating a rebuttal to any 103-rejections (obviousness rejections) or in rebutting the threat of invalidation under 35 U.S.C. 103 in a subsequent lawsuit. For example, the inventor should keep a record of sales, market share, and licensing activities. Since commercialization of an invention often does not begin to take place until after the patent application is filed, evidence of commercial success is mainly used to rebut 103-rejections during lawsuits arising after the patent has been issued. Commercial success is an issue that arises only occasionally.

8.2.1.1 Activities in the laboratory

Laboratory notebooks are used in the determination of which of two inventors was the first to conceive of an invention, i.e., to form the idea of an invention and of how it should work. Where Party A and Party B have filed patent applications and where the notebook of Party A reveals the conception of the invention to have taken place on September 8, 1987 and where the notebook of Party B reveals the conception of a near-identical invention to have taken place at an earlier date, e.g., on August 26, 1987, the patent office will reject the application of Party A.

Priority becomes an issue where the Patent Office receives two patent applications claiming the same invention. In this case, the Patent Office decides which of the two patent applications has priority (and should be allowed), and which patent application should be rejected. Priority also may become an issue in a lawsuit between two parties, where each party has a patent, where the claims cover the same invention. In this case, the court decides which of the two patents has priority. Once it is determined which patent has priority, the claim in the patent not having priority is invalidated under 35 U.S.C. 102(g).

Laboratory notebooks are sources of the date of conception, dates revealing continuing and diligent work (or sporadic and interrupted work), dates revealing development of the idea into a working invention, and the date that the invention finally worked. Where the Patent Office determines that two patent applications cover the same invention, it will initiate a procedure called an Interference. During an Interference, the Patent Office may read the relevant laboratory notebooks, and award the priority to the first party to build the working invention, unless the other party can show that it was the first to conceive of the invention, and that it also worked more or less continuously (showed "diligence") in making a working model (in "reducing the invention to practice") [4]. "Diligence" is the "continuous activity toward reduction to practice so that the inventor's conception and reduction to practice are substantially one continuous act" [5]. To be diligent, the inventor does not have to drop all other work and concentrate solely on the new invention [5]. However, in one court case, *Griffith v. Kanamuru* (Fed. Cir. 1987), a lapse in inventive activities for only 3 months resulted in the rejection of the patent application [6]. The importance of the inventor's activities in maintaining a laboratory notebook is revealed below under Case Studies, in *Monsanto Co. v. Mycogen Plant Science Inc.* (D. Del. 1999) [4].

In FDA-regulated environments, such as the quality control laboratory in a drug company, all notebooks are signed and dated page by page. The pages are bound into the notebook with thread, and are serially numbered in advance. The researcher signs each page of the notebook. The FDA also requires that a witness sign each page of the notebook, attesting that he or she has read and understood the material on the page. Inventors interested in patents should also follow the above procedure in order to help convince the court of conception and diligence [7]. Each page of the inventor's notebook may read, "Read and understood by _____ on _____," where the spaces are where the inventor and the witness should cosign and provide the date [8]. The testimony of notebook cosigners (witnesses) may also be important during a lawsuit, where the court wants evidence pertaining to the meaning of a term in the notebook. In *Bosies v. Benedict* (Fed. Cir. 1994), for example, James Benedict's notebook contained the value "n," but did not contain any definition of "n" [9]. Richard Sunberg had signed (witnessed) the notebook entry. Later on, in court, Mr. Sunberg testified that "n" meant a range of values, including one.

Where a laboratory notebook is lost, an inventor may attempt to supply the date of conception by testimony. However, it is not likely that the court will accept this testimony [10, 11]. Where a laboratory notebook is lost or unsigned, a date of conception may be supplied by dates on other documents, such as drafts of the patent application or an Inventor's Disclosure [12].

8.2.1.2 *Activities involving multiple inventors or collaborators*

The activities of inventors include collaboration between various inventors, technicians, and contractors. In consulting with the patent attorney, the inventor should reveal the names of all the persons who contributed to the development of the invention. An interesting fact pattern, illustrating the concept of collaboration between inventors, is revealed in *Kimberly-Clark Corp. v. Proctor & Gamble* (Fed. Cir. 1992), in the Case Studies below [13]. Another fact pattern, showing the example of a consultant who insisted on being named an inventor, but was held *not* to be an inventor, is shown in *Hess v. Advanced Cardiovascular Systems, Inc.* (Fed. Cir. 1997) in the Case Studies [14].

On occasion, an actual inventor may have been omitted from a patent application. If the error in listed inventors had been a simple mistake without deceptive intent, the matter can be resolved by a Certificate of Correction (35 U.S.C. 254). However, if the inventor had been omitted with deceptive intent, the patent will be invalidated [15]. Further details appear in Chapter 6.

8.2.1.3 *Activities involving sales*

Activities of the inventor that can result in the rejection of a patent application include a sale of the invention [3]. The on-sale bar to patenting occurs where the date of the sale or offer to sell the invention occurred prior to 1 year before the filing date of the patent application (35 U.S.C. 102[b]). The interval between the sales event and the filing date is called a 1-year grace period. If the inventor neglects to file the patent application during this grace period, patenting will be forbidden.

On-sale bar analysis is divided into two steps or parts: (1) Did a patentable invention exist at the time of the sale? and (2) Was the "sale" really and truly a sale? If the examiner determines that the answer to the first part is yes and the second part no, then the on-sale bar will not apply. If the examiner determines that the answer to the first part is no and the second part yes, then the on-sale bar will not apply. The on-sale bar applies only when both answers are yes. When faced with a 102(b)-rejection based on the on-sale bar, the inventor will want to persuade the examiner that one or both of these parts does not apply. *Microchemical, Inc. v. Great Plains Chemical Co, Inc.* (Fed. Cir. 1997) illustrates an inventor's successful argument that an invention had only been partially developed at the time of the sale, and that all of the patentable elements had not yet been conceived and described at the time of the sale [16]. The result of this argument was that the on-sale bar was not applied, and the claim in the patent remained valid (see Case Studies).

In re Mahurkar Patent Litigation (N.D. Ill. 1993) and *Mahurkar v. Impra* (Fed. Cir. 1995) illustrate an inventor's successful argument that what looked like a "sale" was not really a sale, and that the "sale" was really a "sham sale" [17, 18]. These two cases are described below in the Case Studies. It has been pointed out that it is by no means settled what type of "sale" will count as a sale for the purpose of activating the on-sale bar [19]. Perhaps the best approach to avoiding application of the on-sale bar is to provide test samples to a collaborator for no charge [20]. Further details appear in Chapter 3.

The following commentary concerns sales, as it applies to 103-rejections. (The following commentary does not concern the on-sale bar.) Companies are pleased when sales are good, especially when a product captures a large share of the market. Documentation of good sales can help the patent attorney in doing his or her job. Specifically, documentation of good sales can help in the rebuttal of 103-rejections. Usually, arguments of a technical nature are used to rebut 103-rejections. However, evidence of commercial success is another tool in the attorney's toolbox of possible arguments that is acceptable by the Patent Office for addressing 103-rejections.

An example of the commercial success argument, which concerned an immunoassay method, involved Hybritech's U.S. Patent No. 4,376,110, as described in *Hybritech v. Monoclonal Antibodies* (Fed. Cir. 1986) [21]. The patent was attacked as being obvious, and it was argued that the relevant claim should be invalidated under 35 U.S.C. 103. Hybritech defended its patent by a number of tactics, including revealing data showing commercial success of its immunoassay method. To quote from the published court case, "sales increased [by] seven million dollars in just over one year, from $6.9 million in 1983 to an estimated $14.5 million in 1984; sales in 1980 were nonexistant.... Hybritech's HCG kit became the market leader with roughly twenty-five percent of the market at the expense of market shares of the other companies." Another example (not discussed) of the commercial success argument can be found in a case involving a patent for screw anchors for stabilizing towers in the ground [22].

Records of skepticism regarding whether the invention will work can also be used to argue against a 103-rejection. Statements of this sort may be published in editorials in trade magazines or in letters to the editor, for example. Although the marketing manager of a company might not be expected to keep a file of published comments that express skepticism of the company's product line, a record of these comments can prove invaluable in rebutting an examiner's rejection under 35 U.S.C. 103 or in preventing invalidation under 35 U.S.C. 103 in a lawsuit.

8.2.1.4 Activities involving public use

Activities of the inventor that can result in the rejection of the patent application include public use of the invention, where the public use occurred prior to 1 year before the filing date of the patent application (35 U.S.C. 102[b]). Public use of an invention starts a 1-year clock ticking. The inventor then has a 1-year grace period to file a patent application. If the inventor misses this deadline, the application will receive a 102(b)-rejection. The inventor may be able to rebut a 102(b)-rejection, if he or she can convince the examiner that on the date of the public use of the "invention," the displayed invention was only a simulated nonfunctional device.

For example, in *General Instrument Corp., Inc. v. Scientific-Atlanta, Inc.* (Fed. Cir. 1993), Scientific Atlanta was accused of displaying its invention, a video signal encryption device, at a trade show [23]. Scientific Atlanta then provided evidence that the device was only a simulated device, and in this way successfully avoided a 102(b)-rejection.

The following concerns the "experimental exception" to the public use bar. Many inventions, such as new drugs administered in clinics and hospitals, road-building materials, and airplanes, cannot easily be experimentally tested in private. Often, testing occurs for greater than 1 year before a patent application is filed with the Patent Office. Where the public use is for experimental purposes, and not for test-marketing purposes, the applicant may try to document that public use is for experimental reasons. A convincing argument regarding experimental use can successfully rebut a 102(b)-rejection. What form does this argument take? Experimental use can be convincingly argued to override the public use bar, if a combination of factors can be shown, as discussed in *Lough v. Brunswick* (Fed. Cir. 1996) [24]. What is important is the totality of these factors. No single factor is absolutely controlling. These factors are:

1. If the inventor requested that the tester sign a pledge of confidentiality.
2. If the inventor periodically monitored the testing work.
3. If the inventor kept records of the testing work.
4. If the inventor supervised the testing work.
5. If the inventor was given back the device after testing was completed.
6. If any payment given to the inventor was only to defray expenses, and not to produce a profit.
7. If the testing was directed to testing features present in the patent claims, then the experimental use exception may be applied. But if testing was directed to testing features not in the patent claims, then the experimental use exception cannot be applied [25].

8.2.2 Exploring invention patentability

To receive a patent, the inventor must file a patent application containing a specification and claims, meet certain deadlines, and pay certain fees. The term "patentability" does not refer to these requirements for receiving a patent. Patentability generally only refers to the ability of the patent application to satisfy three statutes: 35 U.S.C. 101, 35 U.S.C. 102, and 35 U.S.C. 103.

Much of the attorney's or agent's time nurturing the patent application to the point where it is allowed by the Patent Office is spent in reading the examiner's rejections under 35 U.S.C. 102 ("102-rejections") and 35 U.S.C. 103 ("103-rejections"), and in formulating rebuttals or answers to these rejections. Exploring an invention's patentability consists of selecting existing patents and publications ("the prior art") that the examiner is likely to use to perform 102-rejections and 103-rejections, and predicting the rejections that are likely to be made. The attorney then writes the claim of the patent application in a way that sidesteps the devices described in the prior art.

8.2.2.1 An invention may be patented even if a working prototype has not been made

Patenting is quite possible where a prototype or working model of the invention has not yet been made. Where the activities of the inventor involve only armchair science, and where the invention exists only on paper, it is still possible to acquire a patent. A patent to such an invention is called a "paper patent" or a "prophetic patent." What is required in this situation is that the specification contain enough information to allow "one of ordinary skill in the art" to make and use the invention. An example of a patent application where the inventor succeeded in acquiring a paper patent is described below in the Case Studies, in *In re Strahilevitz* (C.C.P.A. 1982) [26].

Patent law requires that the invention be "enabled" at the time of filing the patent application (35 U.S.C. 112). To persuade the examiner in the Patent Office that the invention is enabled, the applicant need only describe the invention in enough detail so that "one of ordinary skill in the art" can make and use the invention. The enablement requirement applies equally to patent applications describing paper inventions and to those describing working versions of the invention.

8.2.2.2   Avoiding 102-rejections and achieving patentability

Where a claim in a patent application receives a 102-rejection, the invention, as described in that claim, is not patentable. The way a 102-rejection works is revealed by the following example. There is only one claim, and this claim contains five "elements," as shown below. The claim describes a refrigerator (Figure 8.1).

I claim a grocer's refrigerator comprising:

1. A condenser;
2. An oil separator;
3. A pump;
4. A reservoir; and
5. An evaporator.

Imagine that an article in *Grocery Equipment Magazine* describes a refrigerator containing the following elements:

1. A condenser;
2. An oil separator;
3. A pump;
4. A reservoir; and
5. An evaporator.

An examiner reviewing the above claim, and aware of the article in *Grocery Equipment Magazine*, will perform a 102-rejection. The inventor may then try to overcome the 102-rejection by pointing out that the article in *Grocery Equipment Magazine* really did not show an oil separator at all, but that it was actually a valve. Alternatively, the inventor may try to overcome a 102-rejection by pointing out that his refrigerator, as described in the specification, also contained an equalizer line (that was not described in *Grocery Equipment Magazine*). Here the inventor would need to amend the claim to

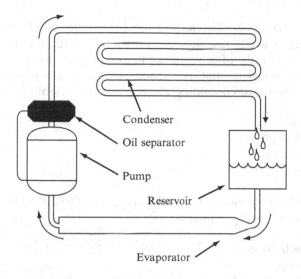

Figure 8.1 Refrigerator. The refrigerator contains a pump, which compresses gaseous refrigerant to produce compressed gaseous refrigerant. The compression process heats the refrigerant. The oil separator removes small amounts of oil that contaminates the warm refrigerant, and returns the collected oil to the pump. The compressed gaseous refrigerant moves in the direction of the arrow to the condenser, where heat is removed, to produce cool, liquid refrigerant. The cool, liquid refrigerant passes to the reservoir, where it is temporarily stored, and then to the evaporator, where useful cooling (cooling of food) occurs. During passage through the evaporator, the refrigerant is converted to a warm gas. The refrigerator may contain other components, such as an equalizer line (not shown).

add an additional element, such as an equalizer line. The amended claim would read as follows:

I claim a grocer's refrigerator comprising:

1. A condenser;
2. An oil separator;
3. A pump;
4. A reservoir;
5. An evaporator; and
6. An equalizer line.

In reading the amended claim, the examiner will notice that the prior art does not describe all of the elements in the claim. The examiner will then allow the claim. The goal of the inventor, of course, is to have the claim allowed. The take-home lesson is that the inventor should reveal the invention to the attorney, so that a claim containing elements 1 to 5 can be put into Claim 1, but that the inventor should also reveal additional novel features that can be included in the claimed invention, such as element 6.

8.2.2.3 *Avoiding 103-rejections and achieving patentability*

The operation of a 103-rejection is like that of a 102-rejection, but more complex. An example of a scenario likely to lead to a 103-rejection is shown below. Imagine that an article in *Grocery Equipment Magazine* describes a refrigerator containing the following elements:

1. A condenser;
2. An oil separator;
3. A pump; and
4. An evaporator;

and that an article appearing in *Lyophilizer Quarterly* describes a refrigerator containing the following elements:

1. A condenser;
2. A pump;
3. A reservoir; and
4. An evaporator.

Now, imagine that the inventor's patent application contains the following claim to a refrigerator (Figure 8.1).

I claim a grocer's refrigerator comprising:

1. A condenser;
2. An oil separator;
3. A pump;
4. A reservoir; and
5. An evaporator.

In this case, the examiner will state that it would be obvious for an ordinary refrigeration engineer to combine the above two references (magazine articles), and come up with the invention as it is described in the claim. In making the statement of obviousness, the examiner is performing a 103-rejection. Most 103-rejections are based on the combination of two or more prior art rejections. However, it is possible to base a 103-rejection on a single prior art reference [27].

A number of strategies can be used to overcome the 103-rejection. First, the attorney can persuade the examiner that he or she failed to understand the article in *Lyophilizer Quarterly*, i.e., that what looked like a reservoir for storing refrigerant was really a barrel of salt. Second, the attorney can try to persuade the examiner that it would *not* have been obvious for an ordinary refrigerator technician to combine the *Grocery Equipment Magazine* reference with the *Lyophilizer Quarterly* reference. For example, the attorney can argue that it would not have been obvious to combine the references because the first reference concerned grocery equipment, while the second reference was in a different field of art and was relevant only to organic chemistry. This type of argument has a special name — it is the argument that "the two references do not occur in the same art" or that "the two references are in non-analogous art." A real-life example of the argument that two references do not occur in the same art is provided by *In re Oetiker* (Fed. Cir. 1992) in the Case Studies [28]. This argument, as it occurred in *In re Oetiker* (Fed. Cir. 1992), succeeded in persuading the examiner to reverse a 103-rejection.

Another method of rebutting 103-rejections is to demonstrate that the claimed invention has an unexpected and surprising property, as compared with the prior art cited by the examiner. While this method of rebutting 103-rejections should be applicable to all technologies, e.g., devices, computers, chemicals, and methods, the following discussion concerns chemicals. In chemical patents, claims covering a molecule can receive a 102-rejection where the examiner finds and cites a publication disclosing the exact, same molecule as that which is claimed. In contrast, 103-rejections can occur where the examiner cites a publication describing a similar or homologous molecule. In performing a 103-rejection, the examiner is required to point out that the two molecules (claimed and prior art) are similar or homologous and also to describe at least one shared property. For example, *In re Dillon* [29] concerned a claim to a tetraorthoester. The claim to the tetraorthoester was rejected in view of a publication describing a triorthoester, where the shared property was use as a water scavenger. Diane Dillon's (the inventor) claim read (in part):

> A composition comprising a hydrocarbon fuel; and a sufficient amount of at least one orthoester so as to reduce the particular emissions from the combustion of the hydrocarbon fuel, wherein the orthoester is of the formula:

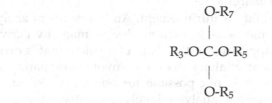

$$
\begin{array}{c}
\text{O-R}_7 \\
| \\
\text{R}_3\text{-O-C-O-R}_5 \\
| \\
\text{O-R}_5
\end{array}
$$

A 103-rejection of this claim can be rebutted by showing that the claimed molecule possesses a surprising and unexpected property. This property can be one that is shared with the prior art molecule, where the property of the claimed molecule occurs at a greater magnitude than shown by the prior art molecule. Alternatively, the unexpected property can be one totally unique to the claimed molecule.

Regarding Diane Dillon's molecule, the examiner cited Elliott U.S. Pat. No. 3,903,006, which described triorthoesters and tetraorthoesters for use as water scavengers. The Elliott patent served to connect triorthoesters and tetraorthoesters by way of a common property, i.e., water scavenger. This connection was sufficient for the 103-rejection, even though Diane Dillon's claim describes emission reduction, and mentions nothing about water scavengers. Diane Dillon argued that the Elliott patent did not show that triorthesters or tetraorthesters, as a fuel additive, could be used for reducing emissions, hoping to fend off the 103-rejection. However, Diane Dillon was not able to show that her tetraorthoester was an unexpectedly better emission reducer than the triorthoester: "she produced no evidence that her compositions possessed properties not possessed by the prior art compositions." In fact, her patent actually contained data showing that the "tri-orthoesters had equivalent activity in reducing particulate emissions." Diane Dillon's claim was thus invalidated under 35 USC 103 by the court.

In re Soni [30] shows a fact pattern that contrasts with that of *In re Dillon*. Pravin Soni claimed a high–molecular weight polyethylene polymer. But the examiner cited patents, where the patents disclosed similar polymers, and Soni's claim received a 103-rejection. Pravin Soni successfully rebutted the rejection by showing that his polymer had unexpected properties: "the improved properties provided by the claimed compositions 'are much greater than would have been predicted given the difference in their molecular weights'." These unexpected and improved properties were disclosed by data in the patent application itself. Thus, Soni's claim remained valid. The fact patterns described in the above cases illustrate that researchers who wish to patent a new chemical should (if possible) tell their attorney or agent of any unexpected and surprising properties of that chemical.

8.2.2.4   Infringment analysis may be conducted at the same time as patentability analysis

An issue related to patentability is infringement. An infringement analysis involves comparing the elements in an existing device, made by Party A, with the written elements appearing in the claim of a patent that is owned by Party B. An infringement analysis does not involve comparing one patent with another patent. It is not possible for one patent to infringe another patent. An infringement analysis involves comparing a device (whether patented or nonpatented) with a patent owned by another party.

Imagine that Party A builds or sells a grocer's refrigerator containing:

1. A condenser;
2. An oil separator;
3. A pump;
4. A reservoir; and
5. An evaporator.

Imagine also that a patent owned by Party B contains a claim that describes or lists these exact elements:

1. A condenser;
2. An oil separator;
3. A pump;
4. A reservoir; and
5. An evaporator.

In the above example, Party A needs to be worried about being sued, because its device is infringing. Party A can avoid infringement by eliminating one of the elements from the device when it is made, for example, by changing its invention so that it contains only these elements:

1. A condenser;
2. A pump;
3. A reservoir; and
4. An evaporator.

A question that sometimes arises is: can an inventor avoid infringement by *adding* an element to the invention, as built? The answer is no. Infringement cannot be avoided by adding new elements to the device [31]. An elegant example (not described here) of infringement analysis, as it applies to recombinant plants, can be found under the heading "2. Literal Infringement of Enzo 931 Patent" in *Enzo v. Calgene*, (D. Del. 1998) [32].

The take-home lesson is that a patentability analysis can guide the inventor in predicting whether his or her invention will pass the scrutiny of the examiner, while an infringement analysis can guide the inventor in predicting if his or her invention, as made, used, and sold, is likely to attract lawsuits.

8.2.2.5   Avoiding 101-rejections and achieving patentablity

Only certain types of inventions are patentable. Claims that describe the wrong type of invention receive a 101-rejection. The types of "inventions" that are likely to be unpatentable under 35 U.S.C. 101 are revealed by a view of 35 U.S.C. 101, as well as by reading certain published cases from the Fed. Cir. and U.S. Supreme Court. These court cases have formulated a number of "exceptions," where these exceptions define inventions that, when claimed, violate 35 U.S.C. 101. These exceptions included printed matter, abstract ideas, and mental acts. At one time, the list of exceptions also included medical procedures and business methods. However, the case law now holds that medical procedures and business methods are patentable. The question of the patentability of computer software has been a difficult one, since software residing in a floppy disk or in a computer memory resembles a book (printed matter), and also because software residing in a floppy disk or in a computer memory resembles a mathematical algorithm (an abstract idea) [33, 34].

Printed matter cannot be patented. The rule stating this is called the "printed matter exception." A goal of the printed matter exception is to channel protection of certain types of creations into patent law, but other creations into copyright law. The printed matter exception applies only where the "printed matter" is not functionally related to the substrate. Music, literary works, and compilations of data generally are not functionally related to the substrate or medium. Substrates include books, floppy disks, CD-ROMs, and computer memories.

Computer programs that provide instructions to a computer and tell the computer how to work provide functionality to a computer. Are all of these types of computer programs patentable under 35 U.S.C. 101? The answer is no. A computer program that provides functionality is not patentable where the program is described in the claim, but where the claim does not describe the medium. The claim must describe the computer program but also the medium, i.e., a floppy disk, CD-ROM, or computer memory [35].

The printed matter exception was applied in *In re Gulack* (C.C.P.A. 1983), and here the claim survived the test [36]. The printed matter that was scrutinized by the court, in this case, consisted of an endless sequence of numbers having mystical qualities. The numbers were printed on a band or ring. Since the nature of the substrate (the ring) was such that it provided an endless character to the numbers (the numbers went around and around), the court held that the printed matter and the substrate were indeed functionally related, and that the claim was patentable. Although the invention was a bit odd, the Gulack decision has had a major impact on patenting in the software industry, and in determining which types of information residing in a floppy disk, CD-ROM, or computer memory are patentable under 35 U.S.C. 101.

A claim that did not survive application of the printed matter exception test can be found in *Bloomstein v. Paramount* (Fed. Cir. 1999) and *Bloomstein v. Paramount* (D. N. Calif. 1998) [37, 38]. The invention concerned a method for altering movies so that the movement of an actor's lips could be changed from speaking a foreign language to movements of lips speaking English. Claim 1 of Bloomstein's U.S. Patent No. 4,827,532 read, "A cinematic work (the printed matter) having an altered facial display made in accordance with a process that includes substituting a second animated facial display...and [where the process is] under the control of...a programmed digital computer." The problem with Bloomstein's claim is that the cinematic work was merely stored on a substrate (a film or laser disc) and read by a projector or computer, and there was no functional interaction. In other words, the images of the lips were not required for the functioning of the movie projector. The claim was held invalid under 35 U.S.C. 101.

The court in *Diamond v. Diehr* (1981) described the "abstract idea exception" [39]. Under this exception, abstract ideas standing alone are invalid under 35 U.S.C. 101. Certain types of mathematical subject matter, standing alone, represent nothing more than abstract ideas [34, 40]. An application of the abstract idea exception appears in *In re Warmerdam* (Fed. Cir. 1994) [41]. Warmerdam's invention related to a bubble hierarchy for collision avoidance between a robot and an object, where a series of bubbles are placed along an axis, and where this axis lies midway between boundary centers of the object. The court pointed out that Warmerdam's Claim 1 was surely to an abstract idea, because in reading Warmerdam's specification, it became clear that the preferred method of using Claim 1 was by a mathematical procedure called the "Hildich method." The court rejected Claim 1 under 35 U.S.C. 101.

In contrast, the court did not reject Warmerdam's Claim 5. Claim 5 described a mathematical algorithm, but it also contained more than a mathematical algorithm — the claim also described a machine — and hence the court did not apply the abstract idea exception to Claim 5. Claim 5 read: "A machine having a memory which contains data representing a bubble hierarchy generated by the method of any of claims 1 through 4." Claim 5 describes a machine in combination with a mathematical formula, and thus could not be invalidated by the abstract idea exception [41].

Where a claim describes an abstract idea, the minimum amount of extra language in a claim needed to impart patentability under 35 U.S.C. 101 is an ongoing unanswered question. The Patent Office has suggested an approach for adding extra language to a claim describing an abstract idea, in order to satisfy 35 U.S.C. 101. This type of extra language has fancifully been called a "safe harbor" [42]. One such safe harbor is the combination of the abstract idea (software program; algorithm; mathematical algorithm) with something physical, such as a robot. Another type of safe harbor occurs where the claim describes signals or data and where the signals or data represent something physical outside the computer, such as signals from a heart or seismic signals.

One problem with including a physical element, or requiring a physical element, as a safe harbor, is that the inventor may really have no interest in patenting any particular physical element [43]. If the programmer has devised a new and useful type of algorithm, should that not be enough? In view of the recent cases of *State Street v. Signature* (Fed. Cir. 1998) and *AT&T v. Excel* (Fed. Cir. 1999), a claim describing an abstract idea (an algorithm alone) may be patentable under 35 U.S.C. 101, where the claim also describes a useful result, even if the useful result is nonphysical [44, 45]. The useful results described in combination with the algorithm were a price for a mutual fund [44] and a method for billing telephone customers at different rates [45]. Please note that prices and billing rates are not physical things.

The following concerns the "medical procedure exception." At one time, methods for diagnosing and treating disease in people were considered to be nonpatentable. This "medical procedure exception" was especially strong before the year 1950 [46]. However, since the 1950s a growing number of patents have been issued that claim medical procedures. These include U.S. Patent No. 4,986,274 (a method for determining the gender of a fetus), U.S. Patent No. 4,874,693 (method for predicting Down syndrome in pregnant women), and U.S. Patent No. 5,080,111 (method for cataract surgery). In order to prevent undue interference with the medical ethics, the Congress recently enacted a law, encoded as 35 U.S.C. 287(c), which prevents lawsuits against doctors who practice the methods that are claimed in patents to medical procedures. 35 U.S.C. 287(c) provides immunity to physicians and other health care providers, such as nurses, dentists, and physical therapists. Note that 35 U.S.C. 287(c) does not provide immunity to the use of a patented drug or instrument. An inventor may apply for a patent to a medical procedure, and be awarded a patent, as the medical procedure exception is no longer used by patent examiners; however, 35 U.S.C. 287(c) will prevent the inventor from suing doctors who use the patented procedure.

Inventions relating to biology can pose unique problems in satisfying 35 U.S.C. 101. Only rarely will a mechanical engineer or electrical engineer design a machine that has no use, and then file a patent application on that invention. However, biologists often file patent applications that claim inventions having no known use. For example, biologists often file patent applications claiming sequences of amino acids (polypeptides) and sequences of DNA, where there is no known biological and pharmacological use. Generally, such claims receive rejections under 35 U.S.C. 101, even if the inventor asserts that the DNA sequence can be used for stimulating further research, or for further laboratory investigation.

The attorney or agent representing a biologist can acquire guidance in claiming polypeptides and nucleic acids from the *Manual of Patent Examining Procedure* (August 2001) [47] and from the *Utility Examination Guidelines* [48]. According to these sources, the attorney needs to show that a newly discovered protein or DNA sequence has three types of utility in order to satisfy 35 U.S.C. 101. These three types of utility are specific utility, substantial utility, and credible utility.

If a newly discovered protein is described in a patent application, and is claimed in that patent application, the claim may be rejected by the patent examiner under 35 U.S.C. 101. If the patent application describes the protein as occurring in liver, intestines, skin, spleen, kidney, T cells, B cells, and dendritic cells, the examiner will explain that the invention lacks specific utility and will do a 101-rejection. But if the inventor can provide data showing that that the newly discovered protein occurs only in intestinal tissue from a patient with severe viral diarrhea, the examiner may be persuaded that the protein has utility in the diagnosis or treatment of viral diarrhea, and may refrain from rejecting the invention due to lack of *specific utility*. If the inventor describes a new protein that occurs only in normal intestinal tissue, then the examiner may still do a 101-rejection, where the

examiner will explain that the invention lacks *substantial utility*. This invention (the newly discovered protein or antibodies to it) will lack substantial
utility, because the inventor will not have any way of connecting the protein
with any real-world disease. But if the inventor can show that the claimed
new protein plays a role in the mechanism of viral diarrhea, then the
examiner may refrain from rejecting the invention due to lack of *substantial
utility*. Here, *substantial utility* can take the form of using large amounts of
the protein for therapeutic purposes, or antibody to that protein for diag-
nostic purposes. If the patent application describes a newly discovered
protein occurring in intestines during viral infection, but if the protein is
coded for by the virus, and the virus is unknown, then the examiner will
reject the claimed invention under 35 U.S.C. 101 because it lacks *credible
utility*. In other words, the examiner will explain that he does not yet believe
that preparation of the protein will be possible, since there does not yet exist
any way to clone it. However, if the inventor can show that he or she has
actually cloned the viral gene, and prepared and purified the new protein,
then the examiner should refrain from rejecting the claim based on lack of
credible utility.

To give another example, if the patent application describes and claims
a transgenic mouse, but where the function of the gene implanted in the
transgenic mouse is unknown, then the examiner will reject the claim
because of a lack of *substantial utility*. If the inventor argues that the transgenic mouse can be used as snake food or as land fill, then the claim will still
receive a 101 rejection, based on lack of *substantial utility*. This is because the
United States Patent and Trademark Office (USPTO) does not accept
"throwaway" utility. To give one more example, claims for drugs that
reverse the aging process or that cure all types of cancer will receive a
101-rejection, based on lack of *credible utility*.

Rejections under 35 U.S.C. 101 are unique in that utility does not have to
exist at the time of filing of the patent application, and does not have to be
described in the patent application, but needs only to be postulated in the
specification. Where a claim receives a 101-rejection, the attorney may
provide new evidence for utility that may not have existed at the time of
filing the patent application claiming a new gene or protein [49]. It is critical
that the specification of the patent application, as filed, contain a hypothesis
relating to the utility. In rebutting the 101-rejection, the attorney should
point out where in the specification this hypothesis is found. New evidence
for utility may be supplied by *research publications* authored by the inventor
or his competitors, by *published patent applications* authored by the
inventor or his competitors, by *unpublished data* such as experimental results
provided by in-house researchers. New evidence appearing in publications
can readily be mailed to the USPTO for consideration by the patent examiner, together with arguments describing how the publications demonstrate
at least one useful function of the claimed invention, e.g., a new gene
or protein. The attorney's argument should conclude with this type of

statement: "The above commentary shows the contemplated invention to have specific, substantial, and credible utility."

New evidence that is not yet published must be accompanied by a Declaration by an expert under Rule 132 (37 C.F.R. 1.132). *Unpublished* evidence for utility may take the form of data from the inventor's own laboratory or data from a patent application filed in a foreign country, or data from a patent application filed with an international patent organization, where that patent application has not yet been published or issued. The Rule 132 Declaration should include a description of the expert's credentials, a step-by-step commentary describing how the claimed invention is expected to work, and a statement that one of ordinary skill in the art would believe the invention to work when provided with the data contained in the Declaration. The Declaration should be accompanied by a copy of the expert's curriculum vitae. To summarize, a 101-rejection may require resumption of communication between the inventor and attorney, where the inventor needs to supply the after-arising data (data made available after the date of filing of the patent application), and where the inventor or one of his or her colleagues needs to sign the Declaration and provide his or her curriculum vitae.

8.3   Information needed to draft a patent application

The inventor needs to provide enough information so that the attorney or agent can describe the invention in enough detail so that "one of ordinary skill in the art" can make and use the invention. Patent law also requires that the patent application reveal the "best mode" of making and using the invention, if work on the invention has progressed far enough so that a "best mode" is known.

The inventor needs to provide the attorney with information on how the invention is distinguished from, or better than, devices or procedures in the prior art. However, the attorney should refrain from including this distinguishing information in the patent application itself, for reasons described below. After the patent application is filed, and where several months go by, and the examiner mails an Office Action to the attorney, the attorney will need the distinguishing information to rebut 102-rejections and 103-rejections contained in the Office Action.

The attorney should refrain from spontaneously writing this distinguishing information in the patent application, or in other documents, such as an Information Disclosure Statement or a Petition to Make Special, as described in *Gentry Gallery v. Berkline Corp.* (Fed. Cir. 1998) [50]. Statements to the effect that one's invention has specific properties that overcome problems found in the prior art can reduce the scope of coverage of the claims, as illustrated by *Dawn Equipment v. Kentucky Farms* (Fed. Cir. 1998) [51], a case concerning farm equipment. Dawn Equipment's device used a rotating pin, while Kentucky Farms' device used an inserted pin and multiple holes. Dawn Equipment, the patent owner, had a good chance of winning the lawsuit, but for one type of comment in its patent. That comment was a series of criticisms of multihole devices of the type made by the competitor, Kentucky Farms. This written criticism of multiholed devices occurring in Dawn Equipment's patent prevented this patent from covering the multiholed device made by Kentucky Farms. A similar lesson was provided by *Signtech USA v. Vutek* (Fed. Cir. 1999) [52], a case concerning ink-jet printers. Another example can be found in *Gentry Gallery v. Berkline* (Fed. Cir. 1998) [50]. The statements written by the attorney during the prosecution of Gentry Gallery's U.S. Patent No. 5,064,244 were that "the invention as described is the only possible configuration of the invention" and that "variations of the described invention are outside the purpose of the invention." Unfortunately for Gentry Gallery, Inc., these statements prevented it from asserting its patent to prevent a competitor from making sofas, where these sofas were variations of Gentry Gallery's sofa.

8.3.1   Examples where not enough information was provided

The inventor needs to tell the patent attorney or agent how to make and use the invention. Where the inventor knows how to make the invention, but willfully refrains from revealing a step (especially a nonroutine step) to the attorney, the patent can be rejected for lack of enablement. Rejections for lack of enablement are under (based on) 35 U.S.C. 112. *Chiron v. Abbott* (D.N.Cal 1995) illustrates the situation of U.S. Patent Application No. 659,339, filed October 10, 1984, where the application failed to provide a description that enables the invention. The application concerned an assay method for AIDS virus [10, 11]. The assay method requires a purified viral protein (envelope protein). Envelope protein is not commercially available. Envelope protein must be produced using a clone of some DNA coding for the envelope protein. The inventor's patent application failed to reveal the source of the cloned envelope protein.

The court pointed out that the inventors had published a paper in the journal *Nature* prior to the filing date, revealing how to clone the envelope gene. The court pointed out that the inventors failed to cite this publication in their patent application. The court pointed out that the inventors had cloned the envelope protein (a good thing), had deposited the cloned envelope protein in a public bank (a good thing), but that the patent application failed to mention this deposit (a bad thing). The publication in *Nature*, as well as the date of deposit, took place before the October 10, 1984 filing date, and hence could easily have been mentioned in the patent application. The claim was invalidated under 35 U.S.C. 112 [10, 11].

The following example concerns an invention of a recombinant plant. Enzo is the owner of U.S. Patent No. 5,190,931. The example is reported

8 *Tom Brody*

in *Enzo v. Calgene* (D. Del. 1998) and *Enzo v. Calgene* (Fed. Cir. 1999) [32, 53]. Dr. Inouye, the inventor in the patent, succeeded in using antisense technology for regulating three genes in prokaryotic cells (in *Escherichia coli*). The inventor's patent application described three working examples from *E. coli*, but failed to mention the inventor's nonworking examples in attempts in eukaryotes (yeast).

The inventor succeeded in persuading the examiner that the antisense method could be applied in eukaryotic cells, and that the method could be accomplished without an unusual degree of experimentation. To persuade the examiner, the inventor filed a Declaration from an independent expert, stating that Dr. Inouye's invention could be applied without undue experimentation to all genes in all cells.

The patent was then issued as U.S. Patent No. 5,190,931. The patent has a claim that reads, "A prokaryotic or eukaryotic cell containing a... DNA construct, which... produces an RNA (antisense RNA) which regulates the function of a gene...."

During a subsequent lawsuit, a number of problems began to unfold for Enzo and its patent:

1. It was revealed that the inventor had stated in a grant application that "there are no consistent ways in which to reliably generate an effective antisense gene for use in the inhibition of a... gene." The grant application was dated April 1, 1990, well after the date of filing the patent application (October 20, 1983). Furthermore, the inventor also stated in an article, published on June 15, 1988, that there was no success with yeast (a eukaryotic cell). Again, this date was after the patent's filing date (October 20, 1983), indicating that these problems also existed at all earlier times;
2. Enzo's patent claimed a method in prokaryotic cells, and contained three examples from prokaryotic cells, but it claimed a method in eukaryotic cells, with no examples; and
3. A witness at the trial stated that trying to make the invention work required "witchcraft" and thus could not have been enabled by the specification.

The above three problems all point to one conclusion — that Enzo's patent application failed to *enable* the invention in eukaryotic cells. Rejections due to lack of enablement will occur where the claims describe an invention where a working model has already been made in the laboratory, but was not adequately described in the patent application. Rejections due to lack of enablement will also occur where the description was overly predictive and where the invention could not possibly have been made in the laboratory at the time of filing.

The following test lends perspective to the above fact pattern. The court applied the Wands factor test [54]. This test involves comparing the following factors:

1. The amount of experimentation required, i.e., whether it is more than a routine amount of experimentation. This factor does not relate to whether an experiment requires an entire year or longer — it relates to whether the time is typical for a worker doing that sort of experiment;
2. The amount of direction or guidance needed;
3. The presence of working examples;
4. The state of the prior art (whether there are many published articles on how to make related inventions work);
5. The predictability of the art — the concept of predictable and unpredictable fields has been discussed [55]; and
6. Whether the claim in the patent claims a great deal more than what is described by the examples in the specification in the patent application, or if the claim only claims what has been shown in the examples.

The court pointed out that Enzo's patent claimed more than what had been described in the working examples, that the claim related to an unpredictable art (biology), and that the working examples were only from *E. coli* (and not from eukaryotes) In short, Enzo's claim failed the Wands test. The court invalidated Enzo's claim under 35 U.S.C. 112.

The take-home lesson is that any step in making an invention that cannot be accomplished by known, routine methods should be described in the patent application to the point where one of ordinary skill in the art can make and use the invention.

8.3.2 *The inventor needs to provide information on the best way (best mode) of making and using the invention*

The patent application must reveal the best way to make and use the invention. Where this requirement is not met, the patent application will be in violation of 35 U.S.C. 112. The best mode is illustrated by *Eli Lilly Corp. v. Barr Laboratories* (Fed. Cir. 2000), in the Case Studies [56]. If a patent application does not reveal the best mode, and the examiner fails to make a 112-rejection and allows the patent, the patent is likely to be invalidated during a subsequent lawsuit. This invalidation will be under 35 U.S.C. 112.

A question related to satisfying the best mode requirement is, what if one of the examples in the specification section does not work at all, i.e., what if the "worst mode" is inadvertently described? The answer is that a patent will probably not be invalidated if one or two of the examples do not work, as long as a greater number of working examples are described [57].

8.3.3   The inventor should provide the attorney with information regarding the level of skill of one of ordinary skill in the field

The inventor needs to provide information on the level of skill of "one of ordinary skill in the art," so that the attorney can write claims that only cover versions of the invention that an ordinary worker can figure out from reading the specification. In other words, the detail in the specification needs to be aligned with the level of skill of an ordinary practitioner in the relevant technology. More detail may be provided, but less detail will invite a 112-rejection. In evaluating patent applications (or patents), the examiner (or court) will use this level of skill as a set point, or balance point, for comparing the "mass" of the specification with the "mass" of the claims.

Amgen v. Hoechst and Transkaryotic Therapies (D. Mass 2001) provides an illustration of the concept of one of ordinary skill in the art [58]. In U.S. Patent No. 5,955,422, Amgen disclosed examples of making the invention (erythropoietin), where one example was with monkey cells and another example was with hamster cells. Examples with other mammalian cells were not shown. Amgen's Claim 1 contained the term "mammalian cells." Transkaryotic Therapies wanted to persuade the court to invalidate the claim, and thus argued that the specification section in Amgen's patent did not contain enough information to guide the ordinary laboratory worker to make the invention in other types of mammalian cells. Transkaryotic Therapies' goal in making this argument was to persuade the court that Amgen's patent should be invalidated for failure to satisfy the enablement requirement (35 U.S.C. 112). Amgen successfully defended its claim by providing an expert, Professor Harvey Lodish, who stated that "it should be easy to figure out experimentally" how to produce erythropoietin in a variety of other mammalian cells.

Another example is mentioned but not discussed. In a lawsuit involving construction materials, the court considered whether one of ordinary skill in the art was "Barney the BrickLayer" or one "familiar with the literature in the field [who] might be expected to supervise building-reinforcement crews and would be likely to attend the World of Concrete Trade Show" [59].

8.4   Reviewing a patent application draft

The draft of the patent application is reviewed in much the same way as a draft for a scientific publication, but with some differences. The draft should be reviewed for accuracy, for typographical errors, and for lapses in logic in the flow of scientific reasoning. If there are figures, the numbers on the figures should match the corresponding numbers in the specification of the patent application. Generally, the specification of the patent application is written in the language of the ordinary scientist or engineer, while the claims are written in terms more familiar to the attorney or agent. However, since it is well recognized that a jury is sometimes the eventual audience of the specification, and also of the claims, the inventor should review both the specification and the claims. A major difference between a scientific manuscript and a patent application is that once an application is filed with the Patent Office, the entry of "last-minute" data is forbidden. This situation is called the prohibition against new matter, under 35 U.S.C. 132.

8.4.1 Example of a patent where the specification succeeded in supporting an invention described in a claim

The inventor needs to review the patent specification to make sure that the descriptions in the claims find support in the specification. Where material in the claims is ambiguous, and the ambiguity cannot be cleared up by viewing the specification, the claim may be invalidated. An example of a fact pattern where the claims contained an ambiguous term, but where the day was saved for the inventor because the specification cleared up the ambiguity, is shown below in the Case Studies in *Amgen v. Hoechst and Transkaryotic Therapies* (D. Mass. 2001) [58].

The specification of the patent application, alone, does not give the patent holder any patent rights, i.e., the right to prevent others from making, using, or selling the patented invention. This right depends on the existence of the claims. The claims, in turn, must be written in a way that carve out the claimed invention from the language in the specification. The inventor needs to provide the patent attorney with enough information so that the attorney can include examples and definitions in the specification, in order to protect the patent, in the event that some ambiguity is found in the claims. Thus, the inventor should read and review the claims before the patent application is filed, to make sure that they are scientifically accurate, and to make sure that they track the supporting material in the specification.

8.4.2 Example of a patent where the specification failed to support an invention described in a claim

Tronzo v. Biomet (Fed. Cir. 1998) provides an interesting example of a patent that could have benefited from a more thorough review by the inventor [60]. Raymond Tronzo's patent, U.S. Patent No. 4,743,262 concerned an artificial hip socket that included a structure called a body. The specification included commentary on a body with a conical shape (a cup with slanted edges). The claims described a body with a conical shape. However, the claims also described bodies having a variety of other shapes, even though the specification described only the conical-type body.

During a subsequent lawsuit, Mr. Tronzo's claims were found to be invalid, because they were too broad, as compared with the specification.

The specification failed to disclose examples of bodies with several different shapes. Biomet, another party, began to manufacture artificial hip sockets with a hemispherical shape. Mr. Tronzo accused Biomet of infringing his patent. The court found that Mr. Tronzo's claim did, in fact, describe Biomet's artificial hip socket, but that Mr. Tronzo's specification did not support the claim. The result was that the court decided that Mr. Tronzo's claim was invalid. The take-home lesson is that the inventor should compare the claims with the specification to make sure that whatever is described in the claims has some corresponding description in the specification.

8.4.3 Writing style in a patent application necessarily differs from the writing style in a scientific publication

In reviewing the patent application, the inventor needs to delete any subjective remarks that may appear in the draft. Scientific publications sometimes contain comments such as "Our discovery simply involved putting two plus two together" or "Our discovery involved combining our earlier published work with the published work of the Japanese research group." Comments like these will inspire the examiner to reject the claims, where the rejections will be under 35 U.S.C. 103 (obviousness rejection). Although an applicant is required under "Rule 56" (37 C.F.R. §1.56) to submit all relevant publications, including publications by the applicant, the applicant is not required to disclose inventive thought processes. Scientific publications also may contain comments as "Our discovery was surprising and amazing" or "The technology we used was difficult and unpredictable." These types of comments, in a patent application, will likely inspire the examiner to allow claims that cover the exact work done by the inventor, but to reject any claims that are broader than the exact work that was done. The purpose of the claims is to cover the invention itself, but also to cover other embodiments, e.g., predicted embodiments, as well as embodiments that cannot be predicted at the time of filing the patent application. Thus, comments as to inventive thought processes, unexpected nature, or difficult nature of the invention should be deleted from any application before it is filed. For example, in *Biogen, Inc. v. Berlex Laboratories* (Fed. Cir. 2003) [62] the patent specification described a method involving one vector, but did not describe a method involving two vectors. The applicants wanted their claims to cover a method involving two vectors. The patent specification contained the subjective comment that there were "technical problems involved with introducing DNA fragments." In view of these subjective comments relating to "technical problems," the court held that the applicant's claims could not cover a method involving two vectors. In addition to proofreading the application for subjective comments, the applicant should also clear up ambiguities in the definitions of words. Specifically, any word appearing in the claims that is to have a meaning other than that found in a

standard dictionary should be clearly defined in the specification — see.
e.g., *Process Control v. HydReclaim* (Fed. Cir. 1999) [63]. The take-home lesson
is that once the patent application is written, the attorney or agent should
give a copy to the inventor, and the inventor should read the claims to make
sure that they are technically accurate, and do not contain contradictions or
errors in logic or grammar.

8.5   Commenting on Office Actions

Office Actions are documents written by the examiner and mailed to the
attorney or agent. An Office Action may contain rejections of all the claims,
allowance of all the claims, or rejections of only some of the claims. Where
there is a rejection, the attorney will ask the inventor if he or she agrees with
the attorney's proposed rebuttals of the examiner's rejections. In acquiring
feedback from the inventor, and in formulating a response to the Office
Action, it is the responsibility of the attorney to refrain as much as possible
from giving up territory that the inventor wants covered by the claims.
Giving up territory can result in a narrower coverage (as expressed in the
language of the claims), but can also prevent the claims from having their
usual penumbra, or aura, of extra coverage. This penumbra of extra coverage extends somewhat beyond the literal language of the claims, and is
provided by the Doctrine of Equivalents.

8.5.1   Prosecution history estoppel

"Prosecution history estoppel" means that communications by the attorney
stop or constrain the interpretation of a word in a claim, almost as though
the content of the communication had been part of the actual claim language. Prosecution history estoppel can reduce the scope of literal coverage
of a claim. For example, if a claim reads, "I claim a pipe connected to a
swivel joint," and if one of the communications from the attorney to the
examiner states that "swivel joint" means only Teflon swivel joints, then
prosecution history estoppel may later prevent the inventor from asserting
his patent to prevent competitors from making pipes connected to copper
swivel joints. Prosecution history estoppel can also reduce the scope of
coverage under the Doctrine of Equivalents, i.e., coverage beyond that
which is literally described in the claims. For example, if a claim reads,
"I claim a pipe connected to a swivel joint," and if a competitor makes a
pipe connected to a universal joint, the patent holder may assert his patent
to prevent the competitor from making the pipe connected to the universal
joint. For this to happen, a court must proclaim that the pipe connected to
the universal joint is equivalent to the pipe connected to a swivel joint.

To understand how prosecution history estoppel can reduce literal
coverage of a claim, there is no need to describe the Doctrine of Equivalents.
However, to understand how prosecution history estoppel can prevent a

slight expansion of claim coverage (beyond that which literally occurs in the claim) under the Doctrine of Equivalents, it is necessary to explain this doctrine.

An example of a coal car patent illustrates the concept of the Doctrine of Equivalents, and how this doctrine is used to expand the coverage of a claim beyond that which is literally written [64]. Ross Winans invented a new type of railroad coal car, where the body (container) was in the shape of a cone, in contrast to the previously used square bin (Figure 8.2). Mr. Winans received U.S. Patent No. 5,175. Mr. Winans' claim read, "What I claim . . . is making the body of a car for the transportation of coal . . . in the form of . . . a cone . . . whereby the force exerted by the weight of the load presses equally in all directions . . . and does not tend to change the form [of the body]." The top view of the body of the patented coal car was round (Figure 8.2). An accused infringer, Mr. Denmead, had manufactured a coal car with a pyramidal body, where the top view was octagonal. The court realized that the accused device was not literally the same as that described in the claim (no literal infringement). However, the court decided that there was an equivalence in the principle of operation and an equivalence in the result, and that the octagonal coal car was equivalent to the patented conical coal car. In both cases, the advantage in operation was provided by the reduced area of the cross section towards the bottom of the body. The court decided that Mr. Denmead was guilty of infringing the claim, because he had manufactured a coal car that was an equivalent to Mr. Winans' patented coal car (Figure 8.2).

The purpose of the Doctrine of Equivalents is to prevent "the unscrupulous copyist to make unimportant and insubstantial changes and substitutions in the patent which, though adding nothing, would be enough to take the copied matter outside [the literal meaning of] the claim. . . ." [65]. The Doctrine of Equivalents is a technique used to expand the scope of a claim beyond the literal words of the claim. It is used during any litigation that may occur after the patent is issued. (The Doctrine of Equivalents is not used during the prosecution phase of a patent, except when the examiner evaluates means + function claims.) It is to the advantage of the inventor (and patent owner) to apply the Doctrine of Equivalents, and it is to the advantage of an accused infringer to argue against the application of the Doctrine of Equivalents.

An accused infringer can utilize the technique of prosecution history estoppel for preventing the application of the Doctrine of Equivalents. "Prosecution history" refers to all the material in the patent application, as it is kept in the USPTO. The prosecution history includes the contents of amendments, Declarations, and written documentation of any oral interviews. The prosecution history includes amendments that are used to comply with an examiner's rejection based on the prior art (102-rejections and 103-rejections), amendments that are used to comply with an examiner's rejection based on 35 U.S.C. 112, amendments that are filed before

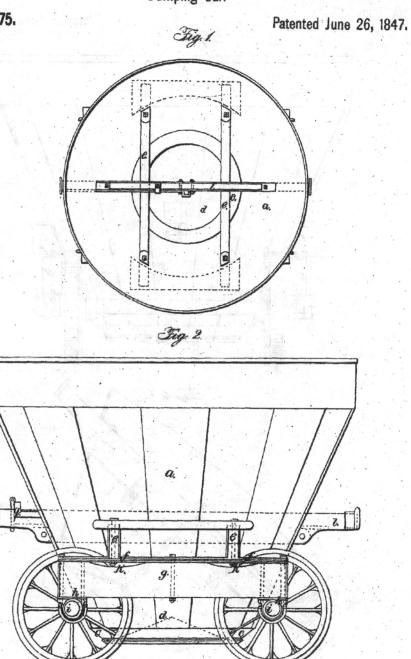

R. WINANS.

Dumping Car.

3 Sheets—Sheet 1.

No. 5,175.

Fig. 1.

Patented June 26, 1847.

Fig 2.

Figure 8.2   (*continued*)

R. WINANS.

Dumping Car.

No. 5,175.

Patented June 26, 1847.

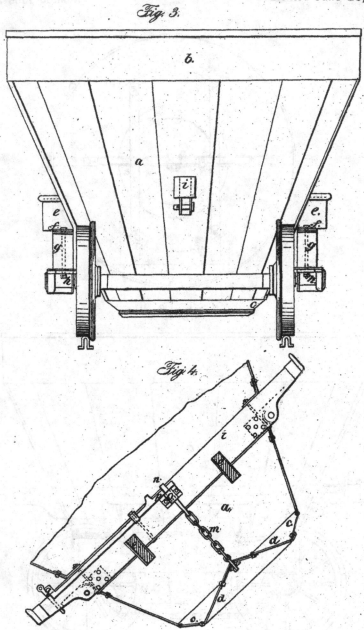

Figure 8.2   (*continued*)

R. WINANS.

Dumping Car.

No. 5,175.

Patented June 26, 1847.

Fig. 6.

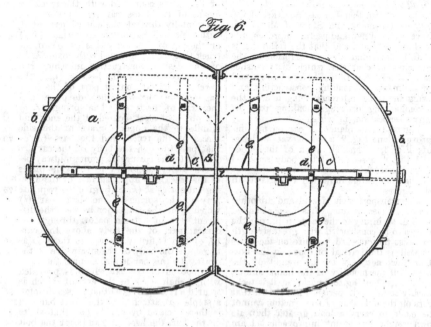

Fig. 5.

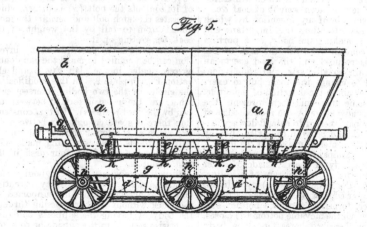

Figure 8.2 *(continued)*

UNITED STATES PATENT OFFICE.

ROSS WINANS, OF BALTIMORE, MARYLAND.

CAR FOR TRANSPORTATION OF COAL, &c.

Specification of Letters Patent No. 5,175, dated June 26, 1847.

To all whom it may concern:

Be it known that I, Ross WINANS, of the city of Baltimore and State of Maryland, have invented new and useful Improvements 5 in Railroad-Cars, and that the following is a full, clear, and exact description of the principle or character which distinguishes it from all other things before known and of the manner of making, constructing, and 10 using the same, reference being had to the accompanying drawings, making part of this specification, in which—

Figure 1 is a plan of a car on my improved plan; Fig. 2, a side elevation there-15 of; Fig. 3, an end elevation of the same; and Fig. 4, a section of the body removed from the truck.

The same letters indicate like parts in all the figures.

20 The transportation of coal and all other heavy articles in lumps has been attended with great injury to the cars—requiring the bodies to be constructed with great strength to resist the outward pressure on the sides 25 as well as the vertical pressure on the bottom, due, not only to the weight of the mass, but the mobility of the lumps among each other, tending to "pack," as it is technically termed. Experience has shown that 30 cars on the old mode of construction cannot be made to carry a load greater than its own weight, but by my improvement I am enabled to make cars of greater durability than those heretofore made which will trans-35 port double their own weight of coal &c.

The principle of my invention by which I am enabled to attain this important end consists in making the body, or a portion thereof, conical, by which the area of the 40 bottom is reduced and the load exerts an equal strain on all parts, and which does not tend to change the form but to exert an equal strain in the direction of the circle. At the same time this form presents the 45 important advantage by the reduced size of the lower part thereof to extend down within the truck and between the axles, thereby lowering the center of gravity of the load.

In the accompanying drawings (a) rep-50 resents the body of the car made of sheet iron in the form of a frustum of a hollow cone, with the upper part (b) cylindrical. To the lower edge of this is secured a flange (c) which forms part of the bottom and 55 against which the movable bottom (d) closes, as will be described hereafter. The body of the car is connected with the truck by means of two cross bars (e, e) that pass horizontally through the conical part of it, with their ends resting on bar springs (f, f) 60 on the top of the side pieces (g, g) of the truck, these being provided with boxes (h, h) of any desired construction in which run the journals of the wheel axles (i, i) the lower end of the conical part of the body 65 passing down between the side pieces and the axles of the truck. The springs (f, f) are plates of steel secured at the ends and middle to the upper surface of the side pieces of the truck, and the bars (e, e) 70 that pass through the body of the car are secured to the springs at points midway between their attachment to the side pieces of the truck, the upper surface of these being cut out as at (k, k) to give the requisite 75 play to the springs. The draft bar (l) which forms the connection between the different cars of a train passes through the conical part of the body above the bars (e, e) and is firmly secured to them so as 80 to relieve the body of the strain due to the draft. To this bar is also secured the movable bottom of the car which is provided with a chain (m) to the end of which is secured a latch piece (n) that passes through 85 a staple (o) attached to the draft bar and is there secured by a bolt (p) that slides on the bar, the head or handle of the bolt being extended outside of the body, as at (q), so that when the car is to be emptied 90 of its contents the bolt (p) is drawn which liberates the latch bolt and permits the movable bottom to fall by the weight of the coal, &c., resting on it.

When desired the principle of my inven-95 tion can be modified to make the car double as represented in the plan Fig. 6 and elevation Fig. 5, plate 2. In this modification the circles of the two bodies intersect each other, and the union is formed between the 100 two by the cord plate (s). In this construction there is space enough left between the two cones at the lower end for the middle pair of wheels, such cars being made with six wheels instead of the four used in the 105 first example.

It will be obvious that car bodies constructed on the principle of my invention may be connected with, and supported on the truck without the bars passing through 110 the body by having the supports bolted or otherwise secured to the outside or to hoops

Figure 8.2   (continued)

2   **5,175**

embracing the whole circumference; but by these modifications a greater strain will be given to the body than by the arrangement above described.

5 What I claim as my invention and desire to secure by Letters Patent is—

1. Making the body of a car for the transportation of coal, &c., in the form of a frustum of a cone, substantially as herein de-
10 scribed, whereby the force exerted by the weight of the load presses equally in all directions and does not tend to change the form thereof, so that every part resists its equal proportion, and by which also the
15 lower part is so reduced as to pass down within the truck frame and between the axles to lower the center of gravity of the load without diminishing the capacity of the car, as described.

2. I also claim extending the body of the 20 car below the connecting pieces of the truck frame, and the line of draft, by passing the connecting bars of the truck frame and the draft bar through the body of the car, substantially as described.

ROSS WINANS.

Witnesses:
GEORGE W. WHISTLER, Jr.,
JOHN B. EASTER.

Figure 8.2 U.S. Patent No. 5,175 concerns a coal car. In the claims, the coal car is described as a "cone." A competitor manufactured a similar coal car having a pyramidal shape. The claims did not literally describe the competitor's pyramidal coal car. However, since the competitor's car was equivalent to that claimed in the patent, the court held that the competitor was guilty of infringement of the patent. When applying the Doctrine of Equivalents, present-day courts usually utilize a formula called the "function, way, result test." Where the device of the accused infringer has an equivalent function, works in an equivalent way, and provides an equivalent result, as the invention described the claims of a patent, then the accused infringer will be guilty of infringing that patent.

any rejection, amendments made for no discernible reason, and repeated assertions of the inventor that are not related to avoiding any rejection. The prosecution history also includes comments that occur in documents such as an Information Disclosure Statement [66] and a Petition to Make Special [50]. Application of prosecution history estoppel is bad for the inventor and patent owner, and good for the accused infringer.

8.5.1.1 Amendment to a claim to avoid a prior art rejection

Inventors may file amendments for various reasons, but the usual reason is to avoid a rejection based on one or more prior art patents or publications. During a subsequent lawsuit, the territory that the inventor has given up by way of the amendment can prevent the inventor from applying the Doctrine of Equivalents to broaden the scope of the claim. *Warner-Jenkinson v. Hilton Davis Chemical Co.*, 520 U.S.17 (1997) provides an example of this situation [67]. U.S. Patent No. 4,560,746, owned by Hilton Davis Chemical Co., claims a method for purifying a dye. Claim 1 reads, "In a process for the purification of a dye ... ultrafiltration through a membrane ... at a pH from approximately 6.0 to 9.0." When the patent application was filed, the claim did not describe the pH limit of 9.0. The examiner rejected the claim in view of Booth's U.S. Patent No. 4,189,380. The Booth patent described an ultrafiltration process that operates at a pH above 9.0. In this patent, Robin Booth and Anthony Cooper described their method in this way: "with polymeric dyes, it has been observed that maintaining the pH of the retentate above 9,

preferably from 11 to 13 ... has an advantageous effect on purification rate."
The inventor of the Hilton Davis application then responded by filing an
amendment that added an upper pH limit of 9.0. During a subsequent
lawsuit, the court held that Hilton Davis was prevented from applying the
Doctrine of Equivalents, with the goal of expanding the scope of its claim to
cover methods operating at above pH 9.0. In layperson's terms, Hilton
Davis was prevented from going back on its word. Hilton Davis' claim
was not permitted to have the extra penumbra of coverage, where this
penumbra would have covered the method at above pH 9.0.

8.5.1.2 Amendment to a claim made for no reason

Sometimes, an inventor files an amendment that changes the wording of a
claim, where the amendment was made for no apparent reason. Sometimes,
amendments are made out of attorney error. During a subsequent lawsuit,
the amendment can prevent the inventor from applying the Doctrine of
Equivalents to broaden the scope of the claim. *Warner-Jenkinson v. Hilton
Davis Chemical Co.*, 520 U.S.17 (1997) provides an example of this fact pattern
[67]. U.S. Patent No. 4,560,746 contains a claim reading, "In a process for the
purification of a dye ... ultrafiltration through a membrane ... at a pH from
approximately 6.0 to 9.0." When the patent application was filed, the claim
did not describe the pH limit of 6.0. (The commentary now concerns the pH
6.0 limit, not the limit at pH 9.0.) Before filing the patent application, Hilton
Davis had tested the method successfully at pH 2.2, i.e., at a pH below 6.0.
However, Hilton Davis by way of an amendment added to the claim the
statement that the method was to be used at a pH of 6.0 or above. There
appeared to be no reason for the addition of this limitation to the claim. The
court held that Hilton Davis was prevented from applying the Doctrine of
Equivalents, with the goal of expanding the scope of its claim to cover
methods operating at below pH 6.0, even though Hilton Davis "offers no
particular explanation as to why the lower level of 6.0 pH [sic] was selected."

8.5.1.3 Amendment to a claim made in order to avoid a 112-rejection

Amendments are commonly filed in order to avoid a 112-rejection. During a
subsequent lawsuit, the amendment can prevent the inventor from applying
the Doctrine of Equivalents to broaden the scope of the claim. In other
words, the court will not allow the inventor to recapture claim coverage
that had been given up when the amendment was made. An example of this
fact pattern is shown below in the Case Studies, in *Bayer AG v. Elan* (N.D. Ga
1999) [68, 69]. In short, the claim described crystals in the dimension range
of 1.0 to 6.0 m^2/g. The examiner rejected the claim, because the specification
did not support crystals of this described range (112-rejection). In response,
the inventor filed an amendment to narrow the range. Prosecution history
estoppel prevented the patent owner from asserting the Doctrine of Equivalents to acquire the usual penumbra of coverage beyond what was literally
written in the claims.

8.5.1.4   Repeated assertions regarding a claim

Repeated assertions made by the inventor during the prosecution of a patent application can result in prosecution history estoppel, even where the assertions are not part of a rebuttal to an examiner's rejection. Bayer Corporation is the owner of U.S. Patent No. 5,264,446 [68, 69]. The patent concerns a crystalline drug for high blood pressure. The patent application claimed crystals in the dimension range of 1.0 to $6.0 \, m^2/g$, as mentioned above. During the course of the prosecution of the patent application, Bayer made repeated assertions regarding the advantages of the range of 1.0 to $4.0 \, m^2/g$. For example, Bayer submitted a Declaration from Dr. Porges, showing the unexpected properties of crystals in the range of 1.0 to $4.0 \, m^2/g$. In a later argument, Bayer wrote that the invention concerns "a special form of nifedipine, namely, having a surface area of 1.0 to $4.0 \, m^2/g$." At a still later point, Bayer provided test data regarding the 1.0 to $4.0 \, m^2/g$ crystals, but did not provide data in the range beyond $4.0 \, m^2/g$. As mentioned above, Elan (the accused infringer) manufactured the same drug, where the crystal dimensions were $6.15 \, m^2/g$. Although it was obvious that Elan's crystals were not literally in the range described in Bayer's claim, Bayer argued that the Doctrine of Equivalents could be applied to expand the range of its claim to cover crystals of the dimension $6.15 \, m^2/g$. However, the court refused Bayer's argument and pointed out that Bayer's repeated assertions (in the prosecution history), regarding the range of 1.0 to $4.0 \, m^2/g$, stopped Bayer from expanding its claim coverage in this way.

8.5.1.5   Details of types of prosecution history that can later prevent the application of the doctrine of equivalents

Classical prosecution history estoppel involves the following series of steps:

1. The inventor files a patent application containing one or more claims. Each claim is composed of several elements or parts. During any lawsuit that may later occur, each element may be surrounded by a penumbra representing coverage via the Doctrine of Equivalents;
2. The examiner cites prior art patents and makes a 102-rejection or a 103-rejection to reject a claim;
3. In response to the rejection, the inventor files an amendment that adds or changes a feature to the claim, thus making the claim novel and different, when compared to the cited prior art. The additional feature results in the amended claim having a narrower scope of claim coverage (covering a more specialized group of devices) than the claim as originally filed;
4. The patent is allowed and it is issued as a U.S. Patent;
5. In an eventual lawsuit, an accused infringer builds a device that is the same as that described in the inventor's claim, as originally filed;

6. The inventor tries to argue that the application of the Doctrine of Equivalents to the amended form of the claim results in the claim's covering the accused device; and
7. The court applies prosecution history estoppel, and the inventor is stopped from making the above argument.

An inventor will be stopped from making the above argument (Step 6) where the examiner had performed a 112-rejection (Step 2). This type of estoppel is called "enablement estoppel" [70].

As discussed in *Festo v. Shoketsu* (Fed. Cir. 2000), an inventor is stopped from making the above argument (Step 6) where the inventor filed an amendment (Step 3) before the examiner had a chance to mail a rejection (Step 2) to the inventor [71, 72].

An inventor is stopped from making the above argument (Step 6) where the inventor had filed an amendment (Step 3) in the total absence of any examiner's rejection (Step 2) [71].

An inventor is stopped from making the above argument (Step 6) where the inventor filed the amendment (Step 3) in the total absence of any examiner's rejection (Step 2), and where the reason for the inventor's amendment is not known and is a total mystery [67].

Once an amendment is made, the inventor is prevented from utilizing the Doctrine of Equivalents to recapture all the claim coverage that had been given up by the amendment, but is also prevented from utilizing the Doctrine of Equivalents to partially recapture the claim coverage. In other words, the claim element that had been amended is no longer surrounded by a penumbra or force field. Application of the Doctrine of Equivalents is modulated by the "flexible bar." Flexible bar refers to an additional doctrine that bars application of the Doctrine of Equivalents under certain conditions, but is more flexible and doesn't bar the application of the Doctrine under other conditions (where the desired equivalent is only tangentially related to what had been given up by the narrowing amendment). If the court permits use of the flexible bar, this is good for the patent holder, as it preferred to the more severe total bar. But if the court chooses not to permit use of the flexible bar, this could be bad for the patent holder and good for the infringer.

If an applicant had submitted a claim reading "rubber or any other elastic material" but later amended it to read "rubber," the court will prevent application of the Doctrine of Equivalents to have an aura that covers other elastic materials, e.g., elastic plastics or elastic metals.

Under the total bar (non-flexible bar), the court will also prevent application of the Doctrine of Equivalents to allow the term "rubber" to have an aura that covers non-elastic materials such as concrete. But under the flexible bar the court will prevent the term "rubber" from covering non-rubber elastic materials, but will permit application of the Doctrine to allow the term "rubber" to cover non-elastic materials, such as concrete. "Non-elastic materials" was only tangentially related to what had been given up. The

U.S. Supreme Court recently held that the flexible bar is, and remains, a valid tool for breathing life into the Doctrine of Equivalents, even where a claim element had been amended [72].

A description of the "flexible bar" to the Doctrine of Equivalents is described in *Goodyear v. Davis* (1880) [73]:

1. An inventor filed a patent application for dentures. The claim, as filed, read:
   "forming the plate and gums in which the teeth are inserted in one piece of rubber, or vulcanite, i.e., an other elastic material."
2. During prosecution, the inventor deleted "other" and added "hard." The result is that the claim no longer alluded to other elastic materials. The amended claim read:
   "forming the plate and gums in which the teeth are inserted in one piece of hard rubber, or vulcanite, i.e., an elastic material."

The overall picture is that the amendment resulted in the claim giving up coverage of dentures made of any kind of *elastic material*. The court did not allow the inventor to use the Doctrine of Equivalents to recapture dentures made of all types of *elastic materials*, as these had been given up by the amendment. However, the court did allow the amended claim to cover some range of equivalents, and allowed the inventor's claim (in its final, amended form) to cover various types of *nonelastic materials*.

Another example of the flexible bar to using the Doctrine of Equivalents is described in *Creo Products v. Presstek* (D. Del, 2001) [74]. A patent application originally contained a claim, claiming a motor that could rotate at a rate of at least 50 r.p.m. During prosecution, the attorney discovered a prior art publication describing a similar motor that could rotate only at one rate (50 r.p.m.). To avoid this prior art, the attorney changed the claim to read "rotation at only 25 r.p.m." In other words, the scope of claim coverage retreated and gave up coverage of motors rotating at rates above 25 r.p.m. Assuming that prosecution history estoppel did not exist, the inventor might still be able to apply the Doctrine of Equivalents to cover motors rotating at 20 to 30 r.p.m. (since 20 to 30 r.p.m. is not quite 50 r.p.m.). Under the flexible bar to applying the Doctrine of Equivalents, the claim might be able to cover motors rotating anywhere in the range of 15 to 20 r.p.m. But under the inflexible bar that prevents application of the Doctrine of Equivalents, the claim will only be able to cover motors that can rotate at exactly 25 r.p.m.

8.5.1.6 Strategies

It is to the advantage of the patent owner to use claim language that will allow him or her to apply the Doctrine of Equivalents, and to avoid claim language that will lead to prosecution history estoppel. Prosecution history estoppel may be avoided by the following techniques:

- Try to refrain from making spontaneous amendments. Refrain from making amendments in response to examiner's rejections, and instead submit more carefully reasoned arguments to the examiner rebutting the rejections.
- Write the claims in a way that intentionally avoids, or skirts around, the prior art. In this way, one can reduce prior art rejections from the examiner and avoid any need to amend the claims.
- Write claims in the means + function form, in addition to the conventional form [75, 76, 77]. Claims in the means + function form contain their own range of equivalents. This range of equivalents, codified in 35 U.S.C. 112, paragraph six, is carefully described and explained in *Kemco v. Control Papers* (Fed. Cir. 2000) [77]. The range of equivalents of a means + function claim exists independently of that which is provided by the Doctrine of Equivalents. The range of equivalents of a means + function claim has not been curtailed by the Festo decision.
- Since prosecution history estoppel applies to a specific element of a claim, and does not bleed over to different elements within the claim, the attorney or agent should clearly divide the claim into various elements [75]. Do not write a claim like this: "A widget comprised of a metal pipe having an assembly at one end, said assembly comprised of a swivel joint and a venturi." If "swivel joint" is later amended to read "snap connector," the court will not allow any range of equivalents to "snap connector," but might also not allow any range of coverage to "venturi." Instead write claims like this: "A widget comprised of (a) A metal pipe; (b) A swivel joint; and (c) A venturi. Here, if "swivel joint" is amended to read "snap connector," the court will forbid any range of coverage to snap connector element, but will not forbid application of the Doctrine of Equivalents to the venturi element.
- File a patent application containing many independent claims, each with a relatively narrow scope of coverage, rather than filing a patent application containing only a few independent claims, each with a relatively broad coverage [75].

8.5.2   Use of declarations

Declarations are documents that are filed (submitted; mailed) with the Patent Office. Declarations are filed for a variety of reasons, as revealed below. Imagine that Party A files a patent application. A Declaration under Rule 131 (37 C.F.R. 131) can be used to overcome an examiner's rejection, where the rejection is based on a prior art reference that only describes Party A's invention but does not claim Party A's invention. A shorthand way to describe this situation is: "The prior art describes, but does not claim, the invention." How can a patent describe but not claim an invention? This situation occurs where the invention is described in the specification but not in the claims. A Rule 131 Declaration cannot be used to overcome an examiner's rejection where the rejection is based on a prior art reference that describes Party A's invention *and* claims Party A's invention. A short-hand way to describe this situation is: "The prior art reference describes and claims the invention." Where the examiner uses a prior art reference to reject a claim in a patent application, where the prior art reference describes *and* claims the invention, and if the inventor feels that he or she was the first to conceive the invention, the inventor cannot file a Rule 131 Declaration. Instead, the inventor must use a procedure known as an Interference (35 U.S.C. 135; 37 C.F.R. 1.601-1.690).

Where an inventor files a Rule 131 Declaration, and the declaration is successful, the result is that the prior art patent reference remains a valid patent. But in an interference, where an inventor files a declaration that successfully establishes priority, the prior art patent reference can be invalidated.

8.5.2.1 Declarations to establish the earlier date of making or building the invention

Declarations are often used for establishing the date of conception of an invention, or the date of first building a working model. *In re Asahi America* (Fed. Cir. 1995) illustrates a fact pattern where the inventor succeeded (just barely) in persuading the examiner of the date [78]. An inventor, Christopher Ziu, filed a patent application on October 20, 1988. Mr. Ziu's invention was a nested pipe, where the inner pipe carried a caustic fluid and the outer pipe served as a container to retain leaks. The examiner rejected Ziu's patent application because another patent application (the Sweeney application) described the same nested pipe invention, and because the Sweeney application had an earlier filing date (April 9, 1987). To overcome the examiner's rejection, Ziu needed to prove to the examiner that he had actually built a working version of the nested pipe on a date before April 9, 1987. Thus, Ziu prepared a Declaration under Rule 131 supplying a series of facts. The Declaration revealed the existence of a published article, which described Ziu's actual invention (not somebody else's invention), where the publication date was before April 9, 1987. The court decided that the publication adequately described Ziu's own working invention. Hence, Ziu was able to establish that his date of building the invention occurred prior to Sweeney's filing date. It was fortunate for Mr. Ziu that he was able to establish a date of building a working model, by providing such documentation from an article in a trade journal. However, the best tactic is to document inventive activities and dates in a laboratory notebook, and not rely on chance documentation of your invention by a journalist in a trade magazine.

A clear-cut example of a declaration used to establish priority in making an invention, but where the declaration failed because of unconvincing data, is shown in the Case Studies in *Schendel v. Curtis, Park, and Cosman* (Fed. Cir. 1996) [79]. The lesson of this case is that declarations should contain convincing data, and not just conclusory statements.

The following is an example of a declaration that failed to convince the court, because it contained fabricated data [80]. Jack Herman thought of an idea of breathable waterproof sock. He filed a patent application on March 31, 1982. The examiner rejected Mr. Herman's patent application because of the existence of U.S. Patent No. 4,344,000, which had been filed at an earlier date of April 22, 1980. Mr. Herman wished to overcome this rejection. In an attempt to overcome the rejection, he filed a Declaration (37 C.F.R. 131) stating that he had made a working version of the waterproof sock before April 22, 1980. The Declaration read, "By January 26, 1980 we constructed ... prototypes from the material provided by Gore (Gore-Tex). I tested them for waterproofness by wearing them in the shower and found that they kept my feet dry."

In a subsequent lawsuit, the court decided that these were false statements, and that the statements were attempts to deceive the Patent Office. The court then invalidated Mr. Herman's patent. The take-home lesson is that one should not willfully make false statements to the Patent Office. The above story has an epilogue. In a later-occurring court case, Mr. Herman argued that his own declaration was incorrect because of human error and that he had no intent to deceive the Patent Office. Here, the court reversed the previous court's decision of invalidity [81]. The take-home lesson is that one should try to refrain from submitting documents to the Patent Office that are likely to raise the issue of impropriety. Indeed, an overall goal of the patent attorney or agent is to write patent applications, and related documents, in a way that raises as few issues as possible.

8.5.2.2 Declarations to establish the skill level of "one of ordinary skill in the art"

The inventor may need to establish the skill of "one of ordinary skill in the art" in a number of circumstances. For example, an examiner may reject (112-rejection) a patent application because the specification does not have sufficient information to enable one of ordinary skill in the art to make the invention. The inventor can then submit a declaration to try to convince the examiner that the specification, in fact, is enabling to one of ordinary skill in the art, and that the rejection should be reversed. An example of this fact pattern appears below in the Case Studies in *Refac v. Lotus* (D. S. New York 1995) [82].

8.5.2.3 Declarations to persuade the examiner that a prior art patent is invalid and should not be used in a prior art rejection

In evaluating a patent application, the examiner compares the invention as described in the applicant's claims with devices (made by the inventor or by a competitor) described in the prior art. If the prior art (either alone or in combination with another) reveals the same invention as that described in the claims, the examiner will reject the claim. If the inventor feels that the examiner is wrong, because of some misunderstanding of the existing patent or publication, the inventor can simply supply the examiner with a conventional rebuttal. On occasion, however, the inventor may believe that the competitor's device or method, as described in the patent or publication, does not work. In this case, the inventor may file a declaration containing data demonstrating that the device does not work, as revealed in the example of *Promega v. Novagen* (D. W.Wisc. 1997), appearing below in the Case studies [83].

8.6   Case studies

The following case studies are cited above, and are intended to be woven into the above commentary.

8.6.1   *Amgen v. Hoechst and Transkaryotic Therapies* (D. Mass. 2001)

This case reveals the fact pattern where an ambiguity in the claims was cleared up by viewing the specification. Amgen's U.S. Patent No. 5,955,422 concerns a protein called erythropoietin. The specification describes the protein, how to produce it in growing cells, and how to purify it. Amgen's claim reads: "Claim 1. A pharmaceutical composition comprising...human erythropoietin...wherein said erythropoietin is purified from mammalian cells grown in culture."

As it turned out, another company, Transkaryotic Therapies (TKT) had produced their own version of erythropoietin, where the erythropoeitin was collected from the fluid (the culture medium) that bathed and nurtured the cells, rather than from the cells themselves. Amgen accused TKT of infringing Claim 1 of the patent. TKT asserted that it was innocent, and argued that the language in Amgen's claim, "purified from mammalian cells grown in culture," does not mean "purified from the culture medium." Fortunately for Amgen, the specification of the patent contained a number of examples of how to make the claimed invention. Example 10 described a technique for obtaining purified erythropoietin from the cell culture medium. The court took note of Example 10, used this example to help interpret the phrase "purified from mammalian cells grown in culture," and decided that this phrase included, in its meaning, "purified from the culture medium surrounding the cells." In short, the inventor's foresight in supplying the patent attorney with Example 10 proved invaluable in persuading the court that TKT was an infringer. The take-home lesson is that the use of several examples in the specification can help avoid problems, due to ambiguities in the claim language.

8.6.2   Bayer AG v. Elan (N.D. Ga 1999); Bayer AG v. Elan (Fed. Cir. 2000)

This case reveals how an amendment made by the attorney (during patent prosecution) can prevent application of the Doctrine of Equivalents (during a lawsuit). Bayer is the owner of U.S. Patent No. 5,264,446. The patent concerns a crystalline drug for high blood pressure. The claim, as originally filed, described crystals in the dimension range of 0.5 to $6.0\,\mathrm{m^2/g}$. The examiner rejected the claim (112-rejection) because this range was not described in the specification. To address the examiner's rejection, Bayer filed an amendment requesting that the range be changed from $1.0{-}6.0\,\mathrm{m^2/g}$ to the narrower range of $1.0{-}4.0\,\mathrm{m^2/g}$. Elan, the accused infringer, manufactured the same drug, where the crystal dimensions were described as greater than $5.0\,\mathrm{m^2/g}$. Although it was obvious that Elan's crystals were not literally in the range described in Bayer's amended claim, Bayer argued that the Doctrine of Equivalents could be applied to expand the range of its claim to cover crystals of about $5.0\,\mathrm{m^2/g}$ or beyond. Bayer wanted to recapture the range that was described in the claim, as originally filed. However, the court pointed out that Bayer's amendment, used to overcome the 112-rejection, stopped Bayer from expanding its claim coverage in this way.

8.6.3   Eli Lilly Corp. v. Barr Laboratories (Fed. Cir. 2000)

This case concerns the requirement for describing the best mode for making and using the invention. The patent, U.S. Patent No. 4,314,081, was owned by Eli Lilly Corporation. It claims an antidepressant drug, fluoxetine (Prozac®). Claim 5 of this patent is directed to the chemical compound. Barr Laboratories (Barr) accused Eli Lilly of failing to reveal the "best mode" in the specification. Barr wanted the court to invalidate Eli Lilly's patent because of the apparent failure to reveal the best mode of practicing the invention. Specifically, Barr pointed out that Eli Lilly had failed to describe the best way of manufacturing the starting material (trifluoromethylphenol) in the pathway of organic synthesis of the drug. The situation was that trifluoromethylphenol was commercially available at the time that Eli Lilly filed its patent application with the Patent Office, but that Eli Lilly had devised a newer, cheaper method of making the starting material, and had not described the cheap method in the patent application. The question before the court was, "Should Eli Lilly's failure to describe the best way (the cheaper way) to prepare the starting material invalidate the claim?"

The court decided that Eli Lilly had satisfied the best mode requirement because:

1. Eli Lilly's patent contained a claim directed to fluoxetine, not to trifluoromethylphenol;

2. The starting material was commercially available at the time the patent was submitted to the Patent Office;
3. Cost and quantity of the available starting materials are irrelevant to determining if the claim satisfies the best mode requirement; and
4. All Eli Lilly was required to do was to name the starting material, providing that it was readily available or synthesizable, and Eli Lilly had in fact named this compound.

The take-home lesson is that where a material or chemical is commercially available, or can be made by one of ordinary skill in the art using publicly available information, there is no need to describe how to make the material or chemical. The concept that features of an invention useful for commercialization can or cannot influence patentability are also discussed in cases relating to polyester resin [89] and a housing for an interferometer [90].

8.6.4 Hess v. Advanced Cardiovascular Systems Inc. (Fed. Cir. 1997)

This case concerns inventive activities involving multiple collaborators, and reveals a problem in inventorship. The case involved two inventors, Dr. Simpson and Dr. Robert, who developed a balloon catheter. The inventors tried various materials (PVC and Teflon) for making the balloon, but these materials did not work, and balloon expansion could not be properly controlled. The inventors then consulted with Mr. Hess, an engineer at Raychem Corp. Mr. Hess suggested use of a heat-shrinkable irradiated polyolefin tubing, and a seal that did not require adhesives. Following this consultation, the two inventors worked alone (without Mr. Hess' participation). At one point, Mr. Hess stated that his suggestion was published in textbooks, and that the suggested process was generally known. Drs. Simpson and Robert filed their patent application. Mr. Hess filed a lawsuit to be named as a coinventor. The court decided that Mr. Hess was only supplying generally known information, that he was only acting as a sales representative, that he did not supply continuing input to the inventive process, and concluded that Mr. Hess was not a coinventor. (If the court had decided that Mr. Hess should have been named as a coinventor, the patent could have been invalidated, because of the failure to name him as an inventor.) The lesson is that consultants who supply only information that is generally known, will probably not be considered to be inventors.

8.6.5 Kimberly-Clark Corp. v. Proctor & Gamble (Fed. Cir. 1992)

This case concerns inventive activities where the issue was whether there were multiple inventors. The case describes an invention of a new diaper, where the diaper contained elasticized flaps to prevent runny leakage. The diaper flaps had been invented by Mr. Lawson at Procter & Gamble (P & G), and resulted in a patent application and eventually in a patent. Mr. Lawson worked alone and was totally unaware of earlier work (resulting in the same invention) by another employee (Mr. Buell) at P & G. After Mr. Lawson's patent application was filed, P & G wished to add Mr. Buell as an inventor, because his inventive activities had occurred on an earlier date than Mr. Lawson's. However, the court refused this request to allow Mr. Buell to be listed as an inventor. The reason was that even though Mr. Lawson and the other inventor worked for the same company, they did not collaborate. The take-home lesson is that there is no requirement that joint inventors must physically work together, but there is a requirement for some form of collaboration (35 U.S.C. 116).

8.6.6   In re Mahurkar Patent Litigation (N.D. Ill. 1993); Mahurkar v. Impra (Fed. Cir. 1995)

This case concerns the on-sale bar. The on-sale bar has its basis in 35 U.S.C. 102(b). Mr. Mahurkar is the inventor in U.S. Patent No. 4,583,968. The inventor argued that there was no sale, and that the on-sale bar did not apply. Another party, accused of infringing this patent, argued that the on-sale bar did apply, and that the patent should therefore be invalidated under 35 U.S.C. 102(b). The inventor agreed that something resembling a sale occurred prior to 1 year before the filing of the patent application, but that it was really a "sham sale." The court agreed that it was a "sham sale" and refused to invalidate U.S. Patent No. 4,583,968. The inventor provided the court with five reasons (see below), where these reasons succeeded in persuading the court that there was really no sale. Mr. Mahurkar had a licensing agreement with Wayne Quinton, owner of an instruments company. Mr. Quinton cherished the licensing agreement and did not wish to give it up. The licensing agreement stated that there must be a sale by a specific date in order for the licensing agreement to remain in effect. The invention (catheter) had been built, and all of the patentable features were in the catheter. However, the catheter was only a crude working model, and could not be commercially sold. So Mr. Quinton devised a "sham sale" in order to keep the licensing agreement alive, but at the same time, not activate the on-sale bar. During a subsequent lawsuit, the court was persuaded that the "sale" was not enough of a sale to activate the on-sale bar, because:

1. The sales agreement, which involved Mr. Quinton and Northwest Kidney Center, stated that the catheter could not be used with human patients;
2. The catheter was not sent to the shipping department of the Northwest Kidney Center, but stored in a little-known cabinet;

3. Mr. Quinton purposefully included false instructions for using the catheter that, if followed, would immediately melt the catheter into a mass of useless plastic;
4. The Northwest Kidney Center never used or needed the type of catheter that was provided; and
5. The catheter contained all the patentable features, but was hand-made and contained rough edges that would cause hemolysis, and dead spots that would cause blood clots, and thus harm any human patient.

Bard, Inc. v. M3 Systems Inc (Fed. Cir. 1998) illustrates another case that concerned whether a "sale" was really a sale [84]. Here, it was held that where the inventor can argue that the "sale" was actually a cost defrayment between collaborators, there was really no sale, and the on-sale bar did not apply.

8.6.7 *Microchemical, Inc. v. Great Plains Chemical Co, Inc.* (Fed. Cir. 1997)

This case concerns one of the two steps required to activate the on-sale bar, i.e., the step that the invention be ready for patenting at the time of the sale. Readiness for patenting can be demonstrated by proof that, at the time of the sale, there had existed descriptions that were sufficient to allow one of ordinary skill in the art to make and use the invention. Readiness for patenting can also be shown by proof that a working model of the invention had been made at the time of the sale [85]. *Microchemical v. Great Plains Chemical* concerns one of the two steps required to activate the on-sale bar, i.e., the step that the invention be patentably complete, or described in enough detail to allow one of ordinary skill in the art to make and use the invention, at the time of the sale. *Microchemical v. Great Plains* focuses on the date that all of the elements of the invention, and not just one or two of these elements, were conceived and described. The case reveals an argument that the on-sale bar did not apply based on the fact that at the time of the sale, all of the patentable features had not yet been invented. Mr. Pratt had invented a machine for weighing and adding small amounts of ingredients to poultry feed. The machine included bins, weighing hopper, weight frame, mixing vessel, and other components. Mr. Pratt offered to sell the machine to Mr. Isaac, manager of a feedlot located in Kansas, in December 1984. At that time, the machine did not work, because of vibrations that prevented accurate weighing. A month or so later, the inventor thought of ways to acquire accurate weighing (isolation mechanism, pads, stabilizers). The inventor filed a patent application on February 26, 1986. The patent application was allowed, resulting in U.S. Patent No. 4,733,971. During a subsequent lawsuit, the court stated that the following decision tree applied:

If the invention had been complete on the date of the offer to the feedlot manager (thus triggering the start of the 1-year grace period), then the on-sale bar would cause invalidation of the patent, but if the invention had not been complete on the date of offer to the bankers then there could not be an on-sale bar. The court decided that, at the time of the December 1984 offer to sell, the machine had not yet been completely invented, that at that time "there is not yet any invention which could be placed on sale," and that therefore the December 1984 offer to sell did not start the clock ticking for the 1-year grace period. The court refused to apply 35 U.S.C. 102(b) to invalidate Mr. Pratt's claims.

Robotic Vision v. View (Fed. Cir. 2001) also concerns the date that the invention was ready for patenting, but focuses on the requirement that the invention be described in enough detail to allow one of ordinary skill in the art to make and use the invention [86]. The invention was sold before 1 year prior to the date of filing the patent application. At the time of the sale, the invention was complete except that the machine did not contain some routine software. The software had not yet been designed. The inventor argued that the invention had not been completed at the time of the sale, and that the 1-year clock, or grace period, had not started ticking. The court refused this argument because one of ordinary skill in the art (an ordinary programmer) could have made a working version of the machine, given the information that was available at the time of the sale.

It is gratifying to point out the following consistency in the published law case. The standard of completeness of the invention is the same when (1) applying the on-sale bar; and (2) filing a patent application for a paper patent (see below, *In re Strahilevitz* [C.C.P.A. 1982]). In both cases, the standard is that the invention should be described in enough detail to enable one of ordinary skill in the art to make and use the invention.

8.6.8 *Monsanto Co. v. Mycogen Plant Science, Inc.* (D. Del. 1999)

The inventor's activities in maintaining a laboratory notebook, which can help in winning priority disputes, are revealed by this case. Table 8.1 reveals the fact pattern that resulted in the court's decision that the Monsanto patent was invalid. Both patents concerned the same invention. The invention was a plant that produced a bacterial protein, where the production was at high levels, rather than at low levels. The bacterial protein was from *Bacillus thuringiensis*. The gene coding for the bacterial protein was altered by scientists (at Monsanto and at Mycogen) so that its codons would be more compatible with the biochemical machinery in the plant cell. The two patent applications were filed several months apart. Monsanto filed its patent application before Mycogen. Upon examination of the laboratory notebooks from the inventors of both companies, it was determined that Mycogen (August 26) had conceived the invention before Monsanto (September 8).

Table 8.1 Dates of conception, completion of the invention, and filing of Monsanto and Mycogen [4]

	Monsanto	Mycogen
Date of conception	September 8, 1987	August 26, 1987
Date of completion the working invention	August 10, 1988	August 11, 1988 (western blot test)
Date of filing patent with patent office	February 24, 1989	August 7, 1989
Patent that was eventually issued	U.S. Patent No. 5,500,365	U.S. Patent No. 5,380,831
Priority goes to		Mycogen
Patent that was invalidated	Monsanto's patent	

Hence, the court decided that Monsanto's patent was invalid, while Mycogen's remained valid.

Where party A is the first to conceive the invention, this is not quite enough to convince the court that party A has priority over party B. Party A must also demonstrate that it had worked more or less continuously (with diligence) on the invention, from the date of conception to the date of making a working invention. More accurately, the party that is first to conceive the invention (party A) must demonstrate that it was diligent from just prior to party B's date of conception to the date that party A made a working version of the invention. Mycogen succeeded in persuading the court that it showed diligence. Mycogen's laboratory notebooks revealed continuous work (Table 8.2; note that "Bt" means *Bacillus thuringiensis*).

To conclude, the court awarded priority to Mycogen, and invalidated Monsanto's patent. If Mycogen had not been able to show its priority date of August 26, 1987, it is likely that Mycogen's patent would have been invalidated. Furthermore, if Mycogen had not been able to reveal more or less

Table 8.2 Mycogen's continuous laboratory work. Data used to reveal continuous work from just before September 8, 1987 until mid-August 1988 include the following [4].

October 20, 1987	AMVBt2 construct made by Barton
November 2, 1987	AMVBt3 construct made by Barton
January 15, 1988	AMVBt4 construct made by Barton
Mid-January 1988	Transformation experiments conducted by Cannon
May 23–26, 1988	Tobacco hornworm bioassays conducted by Cannon
June 3, 1988	Cannon enters data in her laboratory notebook
June and July 1988	Additional bioassays conducted by Cannon
August 11, 1988	Western blot tests conducted by Miller

continuous work on its invention from just prior to September 8, 1987 to August 11, 1988, it is likely that Mycogen's patent would have been invalidated. It has been pointed out that the date of conception is not the same thing as the date of "clearly articulating the problem." The date of conception is the date that a solution to the problem was provided [87]; the date of clearly articulating the problem has little or no relevance to the patent application process.

8.6.9 In re Oetiker (Fed. Cir. 1992)

This case reveals a technique that can be used to rebut 103-rejections. Hans Oetiker invented a metal clamp, where the clamp contained a hook. The hook maintains the preassembly condition of automobile axles. The examiner rejected Mr. Oetiker's patent application. It was a 103-rejection. The examiner stated that the invention would have been obvious to one viewing two previously existing patents — an earlier patent of Oetiker's (U.S. Patent No. 4,492,004) and Lauro's U.S. Patent No. 3,426,400. The earlier Oetiker patent was in the same field of technology as his present patent application, but did not describe the hook. Lauro's patent was in a different field. It related to ladies' garments, not to automobile axles. Oetiker argued that one of ordinary skill, seeking to solve the problem that was facing him or her, would not look to patents or publications in the field of ladies' garments for a solution. Mr. Oetiker's argument succeeded. The take-home lesson is that when an inventor is faced with a 103-rejection, and if the prior art cited by the examiner is in a different field of art than that of the inventor's application, the inventor should try to utilize Mr. Oetiker's argument.

8.6.10 Promega v. Novagen (D. W. Wisc. 1997)

This case shows the use of a Declaration to establish inoperability of the prior art. The examiner rejected a claim because a similar or identical invention had been described by the "Baranov reference." Promega filed a Declaration revealing some of the work done to show that the Baranov procedure did not work. The Declaration used conventional scientific commentary to convince the examiner that the prior art was inoperative. The Declaration read in part, "The first experiment...utilized the incubation mixture alone as defined in Baranov Example 4. The second experiment... utilized the incubation mixture and buffer A as defined in Baranov Example 4. See Exhibit 1 attached...which shows the components, concentrations, and amount added for 0.5 ml batch Baranov incubation mixture." The examiner allowed the patent application, and it became U.S. Patent No. 5,324,637. The take-home lesson, at this point, is that sometimes an examiner's rejections can be overcome by a Declaration showing that the device or method in a prior art patent or publication does not work.

In a later-occurring lawsuit, there arose a problem with Promega's Declaration. The court stated that the *examiner's understanding* was only heresay, and thus unpersuasive to the court. The take-home lesson is that the inventor should request the examiner to document his or her understanding on paper, and provide copies of this document to the inventor and to the inventor's file in the Patent Office.

8.6.11 Refac v. Lotus (D.S.N.Y. 1995)

This case reveals use of a Declaration to convince the court that the specification contained enough information to allow one of ordinary skill in the art to make and use the invention. U.S. Patent No. 4,398,249 concerns a method of converting software source code to object code. The inventors are Rene Pardo and Remy Landau. After the patent application was filed, the examiner rejected the application, stating that the specification was not sufficient to enable an ordinary programmer to make and use the invention. To convince the examiner that the specification was sufficient, Pardo filed a Declaration (37 C.F.R. 132; Rule 132) stating that the specification was sufficient. The examiner was not persuaded. The examiner responded that declarations by the inventor (rather than by an impartial observer) are "self-serving" and have "very little probative value." The take-home lesson from the case, so far, is that declarations attesting to the ability of the specification of a patent application should preferably be written by an impartial user.

The inventors then filed an additional Declaration from Peter Jones, a computer scientist. Mr. Jones' Declaration stated that, "From the written disclosures and flowchart shown in the drawing of such patent application..." one of ordinary skill in the art could make and use the invention. The examiner was persuaded by this second Declaration, and allowed the patent to be issued.

However, it was later determined that Mr. Jones had previously been a coworker of Pardo in the software company, and concluded that he could not have been impartial when he wrote the Declaration. Furthermore, it was also determined that during Mr. Jones' employment at the software company, he had received instructions on how to use the invention. The court concluded that Mr. Jones was not impartial. The court concluded that Mr. Jones was not of ordinary skill in the art, since he had previously been given special training in the invention. The court held that even though the Declaration contained acts of omission, rather than acts of misrepresentation, the inventors had intentionally misled the Patent Office. The court invalidated U.S. Patent No. 4,398,249. The take-home lesson is that declarations regarding the ability of the specification to enable the invention should be made by an observer who is impartial, i.e., one who has no financial interest in the invention, and should not take liberties with the level of skill of an ordinary practitioner of the relevant technology.

8.6.12 Schendel v. Curtis, Park, and Cosman (Fed. Cir. 1996)

This case illustrates an attempt to establish a date of building a working model. Genetics Institute filed a patent application on August 29, 1990. Immunex filed a patent application on an earlier date, August 22, 1989. The inventions from both companies concerned a fusion protein, i.e., an artificial protein consisting of two natural proteins linked together in one continuous polypeptide. The two natural proteins were interleukin and colony-stimulating factor. Genetics Institute wanted to convince the Patent Office that Immunex's patent was invalid. To do so, a first step was to convince the examiner that Genetics Institute had made a working version of its fusion protein on a date prior to Immunex's filing date (August 22, 1989). The inventor at Genetics Institute wrote a Declaration. The Declaration read, in effect, "I, Dr. Schendel, prepared DNA coding for the fusion protein before August 22, 1989. The protein consisted of interleukin linked to colony-stimulating factor. The fusion protein was tested for biological activity, and it had activity of interleukin and of colony-stimulating factor."

The court was not convinced. The court pointed out that the Declaration failed to provide evidence that the fusion protein existed, since the inventor at Genetics Institute failed to conduct a test to detect the fusion protein. One such test is to sequence the DNA coding for the fusion protein. Another test is to determine the molecular weight of the expressed fusion protein. All the inventor had done was to determine that both of the biological activities were present (the court pointed out that both activities could have been due to a mixture of the two proteins, but in a separate, unlinked state). The result is that Genetics Institute was not able to invalidate Immunex's patent. The take-home lesson is that declarations should contain fact-based information, not just conclusory statements.

8.6.13 In re Strahilevitz (C.C.P.A. 1982)

This case reveals the concept of a "prophetic patent," i.e., where the patent application described how to make and use all of the claimed features of the invention but where no working model had been built. The invention was a device for removing antigens from the blood of a living mammal. The patent described a solid matrix containing immobilized (bound) antibodies. (Antigens are molecules that bind to antibodies.) The method of the invention involved passing blood (containing the antigen) through a filter, and allowing the filtered blood to contact the matrix, where contact with the matrix would remove the antigen from the filtered blood. After processing on the matrix, the processed blood was returned to the living animal. The invention contains an inlet for receiving unprocessed blood, and an outlet for returning processed blood to the animal or human.

The patent examiner rejected the patent application. He complained that the claims claimed application to a large variety of antigens. He complained that the applicant should have provided at least 50 working examples in the patent application, to show that the invention worked. After receiving the rejection from the examiner, the inventor appealed to the court. The court reviewed the reasoning behind the examiner's rejection. The court realized that filters, inlets, outlets, and matrices with immobilized antibodies were all things that operated predictably. The court stated that there was no reason why the 50 different immobilized antibodies to remove the 50 different antigens should work differently from each other, or require different techniques. The court held that the patent application satisfied the enablement requirement of 35 U.S.C. 112, and that the application should not receive a 112-rejection.

The take home lesson is that there is no *per se* requirement that an inventor should wait until the invention is built and tested, before filing a patent application. There is no requirement that the invention be built. All that is required is that enough information be given to allow one of ordinary skill in the art to make and use the invention. A related concept is that the inventor does not have to know or understand how the invention works in order to file a patent application or to receive a patent [87].

The opinions provided in this chapter do not necessarily reflect those of the author's past, present, or future employer.

Acknowledgments

I thank attorneys J. Bruce McCubbrey and Don Bartels for their guidance in infringement and patentability analysis. I thank attorneys David E. Weslow and William C. Rooklidge for answering specific questions regarding the on-sale bar.

References

1. Allison, J.R. and Lemley, M.A., How federal circuit judges vote in patent validity cases, *Florida State University Law Review*, 27, 745–766, 2000.
2. Rooklidge, W.C. and Jensen, S.C., Common sense, simplicity and experimental use negation of the public use and on-sale bars. *John Marshall Law Review*, 29, 1–53, 1995.
3. Brody, T., Sale of your product can prevent patenting: how to avoid the on-sale bar. *Biopharmacology*, 15, 38–42, 2002.
4. *Monsanto Co. v. Mycogen Plant Science Inc.*, 61 F. Supp. 2d 1333 (D. Del. 1999).
5. *Gould v. Schawlow*, 363 F.2d 908 (C.C.P.A. 1966).
6. *Griffith v. Kanamuru*, 2 U.S.P.Q.2d 1361 (Fed. Cir. 1987).

7. Anawalt, H.C. and Enayati, E.F., *IP Strategy Complete Intellectual Property Planning, Access and Protection*, West Group, Minneapolis, 1998, p. 315.
8. *Cooper v. Goldfarb*, 47 U.S.P.Q.2d 1896 (Fed. Cir. 1998).
9. *Elmar Bosies and Rudi Gall v. James J. Benedict and Christopher M. Perkins*, 30 U.S.P.Q.2d 1862 (Fed. Cir. 1994).
10. *Chiron v. Abbott*, 902 F. Supp. 1103 (D. N. Ca. 1995).
11. *Chiron v. Abbott*, 1995 U.S. Dist. LEXIS 8080 (D. N. Ca. 1995).
12. *Genentech v. Celltech*, 2001 U.S. Dist. LEXIS 3489 (N. Ca. March 16, 2001).
13. *Kimberly-Clark Corp. v. Proctor & Gamble*, 23 U.S.P.Q.2d 1921 (Fed. Cir. 1992).
14. *Hess v. Advanced Cardiovascular Systems Inc.*, 41 USPQ2d 1782 (Fed. Cir. 1997).
15. *Trovan, Ltd. v. Sokymat Sa, Irori, and Ake Gustafson*, 2000 U.S. App. LEXIS 22901 (Fed. Cir. 2000).
16. *Microchemical, Inc. v. Great Plains Chemical Co, Inc.*, 41 U.S.P.Q.2d 1238 (Fed. Cir. 1997).
17. *In re Mahurkar Patent Litigation*, 28 U.S.P.Q.2d 1801 (N.D. Ill. 1993).
18. *Mahurkar v. Impra Inc.*, 37 U.S.P.Q.2d 1138 (Fed. Cir. 1995).
19. Rooklidge, W.C. and Hill, R.B., The law of unintended consequences: the on-sale bar after *Pfaff v. Wells Electronics*, *Journal of Patent Trademark Office Society*, 82, 163–180, 2000.
20. Barash, E.H., Experimental uses, patents, and scientific progress, *Northwestern University Law Review*, 91, 667–703, 1997.
21. *Hybritech Inc. v. Monoclonal Antibodies Inc.*, 231 U.S.P.Q. 81 (Fed. Cir. 1986).
22. *Richard Ruiz v. A.B. Chance Co.*, 57 U.S.P.Q.2d 1161 (Fed. Cir. 2000).
23. *General Instrument Corp., Inc. v. Scientific-Atlanta, Inc.*, 27 USPQ2d 1145 (Fed. Cir. 1993).
24. *Lough v. Brunswick*, 39 U.S.P.Q.2d 1100 (Fed. Cir. 1996).
25. *In re Smith*, 218 U.S.P.Q. 976 (Fed. Cir. 1983).
26. *In re Strahilevitz*, 212 U.S.P.Q.561 (C.C.P.A. 1982).
27. *Sibia Neurosciences Inc. v. Cadus Pharmaceutical Corp.*, 55 U.S.P.Q.2d 1927 (Fed. Cir. 2000).
28. *In re Oetiker*, 24 U.S.P.Q.2d 1443 (Fed. Cir. 1992).
29. *In re Dillon*, 16 U.S.P.Q.2d 1897 (Fed. Cir. 1990).
30. *In re Soni*, 34 U.S.P.Q.2d 1684 (Fed. Cir. 1995).
31. *Amstar Corp. v. Envirotech Corp.*, 730 F.2d 1476 (Fed. Cir. 1984).
32. *Enzo Biochem, Inc. v. Calgene, Inc.*, 14 F.Supp.2d 536 (D. Del. 1998).
33. Samuelson, P., Benson revisited: the case against patent protection for algorithms and other computer program-related inventions. *Emory Law Journal*, 39, 1025–1154, 1990.
34. Laurie, R.S. and Siino, J.K., A bridge over troubled waters? Software patentability and the PTO's proposed guidelines, Part I, *The Computer Lawyer*, 12, 6–21, 1995.
35. *Manual of Patent Examining Procedure*, 7th ed., July 1998, revision, Feb. 2000, U.S. Dept. of Commerce, Section 2106, pp. 2100–2112.
36. *In re Gulack*, 217 U.S.P.Q. 401 (Fed. Cir. 1983).
37. *Bloomstein v. Paramount Pictures Inc. and Lucas Digital Ltd.*, 1998 U.S. Dist. LEXIS 20839 (N. D. Ca. 1998).
38. *Bloomstein v. Paramount Pictures Inc. and Lucas Digital Ltd.*, 1999 U.S. App. LEXIS 21391 (Fed. Cir. 1999).
39. *Diamond v. Diehr*, 209 U.S.P.Q. 1 (1981).

40. *In re Alappat,* 31 U.S.P.Q.2d 1545 (Fed. Cir. 1994).
41. *In re Warmerdam,* 31 U.S.P.Q.2d 1754 (Fed. Cir. 1994).
42. *Manual of Patent Examining Procedure,* 7th ed., July 1998, revision, Feb. 2000, U.S. Dept. of Commerce, Section 2106, pp. 2100–2115.
43. Cohen, J.E. and Lemley, M., Patent scope and innovation in the software industry. *California Law Review,* 89, 1–57, 2000.
44. *State Street Bank v. Signature Financial Group,* 47 U.S.P.Q.2d 1596 (Fed. Cir. 1998).
45. *AT&T Corp. v. Excel Communications Inc.,* 50 U.S.P.Q.2d 1447 (Fed. Cir. 1999).
46. Portman, R.M., Legislative restriction on medical and surgical procedure patents removes impediment to medical diagnostics, *University of Baltimore Intellectual Property Journal,* 4, 91–119, 1996.
47. *Manual of Patent Examining Procedure,* 7th ed., July 1998, revision, Feb. 2000, U.S. Dept. of Commerce, Section 2107.
48. *Utility Examination Guidelines,* Federal Register 66, 2001, 1097 — 1099.
49. *Manual of Patent Examining Procedure,* 7th ed., July 1998, revision, Feb. 2000, U.S. Dept. of Commerce, Section 2107.02.
50. *Gentry Gallery Inc. v. Berkline Corp.,* 45 U.S.P.Q.2d 1498 (Fed. Cir. 1998).
51. *Dawn Equipment v. Kentucky Farms,* 46 U.S.P.Q.2d 1109 (Fed. Cir. 1998).
52. *Signtech USA v. Vutek,* 50 U.S.P.Q.2d 1372 (Fed. Cir. 1999).
53. *Enzo Biochem, Inc. v. Calgene, Inc.,* 52 U.S.P.Q.2d 1129 (Fed. Cir. 1999).
54. *In re Wands,* 8 U.S.P.Q.2d 1400 (Fed. Cir. 1988).
55. O'Shaughnessy, B.P., The false inventive genus. *Fordham Intellectual Property, Media, and Entertainment Law Journal,* 7, 147–229, 1996.
56. *Eli Lilly Corp. v. Barr Laboratories,* 55 U.S.P.Q.2d 1609 (Fed. Cir. 2000).
57. *Atlas Powder Co. v. E.I. DuPont,* 224 U.S.P.Q. 409 (Fed. Cir. 1984).
58. *Amgen v. Hoechst and Transkaryotic Therapies,* 57 USPQ2d 1449 (D. Mass. 2001).
59. *Helifix Ltd. v. Blok-Lok Ltd.,* 52 U.S.P.Q.2d 1486 (D. Mass. 1998).
60. *Tronzo v. Biomet Inc.,* 47 U.S.P.Q. 1829 (Fed. Cir. 1998).
61. Romary, J.M. and Michelsohn, A.M., Patent claim interpretation after Markman. *American University Law Review,* 46, 1887–1933, 1997.
62. *Biogen Inc. v. Berlex Laboratories Inc.,* 65 U.S.P.Q.2d 1809 (Fed. Cir. 2003).
63. *Process Control v. HydReclaim,* 52 U.S.P.Q.2d 1029 (Fed. Cir. 1999).
64. *Winans v. Denmead,* 56 U.S. (15 How.) 330 (1854).
65. *Graver Tank and Manufacturing Co. Inc. v. Linde Air Products Co.,* 339 U.S. 605 (1950).
66. *Ekchian v. The Home Depot Inc.,* 41 U.S.P.Q.2d 1364 (Fed. Cir. 1997).
67. *Warner-Jenkinson v. Hilton Davis Chemical Co.,* 520 U.S.17 (1997).
68. *Bayer AG Corp. v. Elan Pharmaceutical Research Corp.,* 643 F. Supp.2d 1295 (N. D. Ga. 1999).
69. *Bayer AG Corp. v. Elan Pharmaceutical Research Corp.,* 54 U.S.P.Q.2d 1710 (Fed. Cir. 2000).
70. Apple, T., Enablement estoppel, *Computer and High Technology Law J,* 13, 107–135, 1997.
71. *Festo Corp. v. Shoketsu Kinzoku Kogyo Kabushiki Co.,* 62 U.S.P.Q.2d 1705 (U.S. 2002).
72. *Festo Corp. v. Shoketsu Kinzoku Kogyo Kabushiki Co.,* 50 U.S.P.Q.2d 1385 (Fed. Cir. 1999).
73. *Goodyear Dental Vulcanite v. Davis,* 102 U.S. 222 (1880).
74. *Creo Products v. Presstek,* 2001 U.S. Dist. LEXIS 6226 (D. Del. 2001).

75. Hosteny, J.N., Does Festo change patent prosecution? *Intellectual Property Today,* 8, 44–45, 2001.
76. *Manual of Patent Examining Procedure,* 7th ed., July 1998, revision, Feb. 2000, U.S. Dept. of Commerce, Section 2186.
77. *Kemco Sales Inc. v. Control Papers Co.,* 54 U.S.P.Q.2d 1308 (Fed. Cir. 2000).
78. *In re Asahi America Inc.,* 33 U.S.P.Q.2d 1921 (Fed. Cir. 1995).
79. *Schendel v. Curtis, Park, and Cosman,* 38 U.S.P.Q.2d 1743 (Fed. Cir. 1996).
80. *Herman v. Brooks Shoe Co.,* 39 U.S.P.Q.2d 1773 (D. S. N.Y. 1996).
81. *Herman v. Brooks Shoe Co.,* 1997 U.S. App. LEXIS 5426 (Fed. Cir. 1997).
82. *Refac International Ltd. v. Lotus Development Corp.,* 38 U.S.P.Q.2d 1653 (D.S. New York 1995).
83. *Promega Corp. v. Novagen Inc.,* 6 F. Supp.2d 1004 (D. W. Wisc. 1997).
84. *Bard Inc. v. M3 Systems Inc.,* 48 U.S.P.Q.2d 1225 (Fed. Cir. 1998).
85. *Pfaff v. Wells Electronics Inc.,* 48 U.S.P.Q.2d 1641 (1998).
86. *Robotic Vision v. View Engineering,* 2001 U.S. App. LEXIS 8479 (Fed. Cir. 2001).
87. *Singh v. Brake,* 55 U.S.P.Q.2d 1673 (Fed. Cir. 2000).
88. *Newman v. Quigg,* 11 USPQ2d 1340 (Fed. Cir. 1989).
89. *In re Rinehart,* 189 U.S.P.Q. 143 (CCPA 1976).
90. *Zygo v. Wyko,* 38 U.S.P.Q.2d 1281 (Fed. Cir. 1996).

Part IV

Ancillary patent activities

chapter nine

How to read a patent

Judith M. Riley
Gifford, Krass, Groh, Sprinkle, Anderson & Citkowski, P.C.

Contents

9.1   Quick tips on reading a patent

- Go first to the cover page
- The Title may or may not be informative
- Focus on the Abstract, a concise description of the invention in broad terms
- Pay particular attention to the drawing selected to appear on the cover page
- Locate what you believe is the broadest claim — often, but not always, this will be Claim 1
- Use the drawings to find the elements of the invention recited in the broadest claim
- Go to the beginning of the Summary for the inventor's characterization of the invention
- Go to the Detailed Description after you have a general understanding of the invention
- Skim through the rest of the claims to discover other aspects of the invention
- Read the Background to find out what problems in the prior art the invention is intended to solve
- Go back to the cover page for nontechnical information about the patent — the names of the inventor(s); when the patent was issued: what entity owns the patent: prior U.S. and foreign applications, what prior art the examiner considered when allowing the claims
- To focus on a key technical term, try a keyword search on an on-line version of the patent

9.2   What? me read a patent?

There is a first time for everything, and welcome or not, the day may come when you have to figure out what particular piece of technology a patent protects. It may be because you are at work on your own invention and need to know whether you are truly inventing or just reinventing. Or perhaps your company has developed a promising new product and wants to be sure that it will not infringe on anyone else's valid patent rights. Or maybe you need to know the extent of the patent rights owned by a potential joint venture partner. Another possibility is that you are simply looking for specific technical information that might appear only in a patent document. These are only a few of the many possible scenarios that may require you to read and interpret a patent.

The task may seem intimidating, but if you take a systematic approach and break it down into a series of ordered steps, even a first-timer can successfully read and understand a patent document. This is so because patents are formal documents, and the information contained in them always appears in the same format. Once you have mastered the format, you should be able to understand the content without undue difficulty. If the technology surrounding the patent is familiar to you, of course it will be that much easier. But you may be surprised at how much you can benefit from reading a patent in a technical field very foreign to you, if you approach the material in a systematic manner.

For those who are skeptical of this last point, consider that patents are written by patent agents and attorneys. While we have special training in drafting patents, our technical backgrounds are not necessarily extensive in the field of a particular patent. Yet, we must be able to translate an inventor's often unclear disclosure into a document that an unknown examiner in the Patent Office will be able to understand and, hopefully, deem worthy of patent protection. We can do this because we break our task down into discrete steps — understanding the invention, searching the prior art and deciding whether it is patentable, framing a set of claims that will adequately protect it, formulating drawings that illustrate its operation, etc. If you approach the task of understanding a patent in the same systematic way, you should succeed with no trouble.

When you look at a patent for the first time, your first reactions may be dismay and confusion — you cannot help but think, where, in all those pages of text and drawings, can I find the information I truly need to know? Do I really need to read the whole document word for word from beginning to end, and if I do not, how can I be sure I will not be missing something important? What in the world is the point of those names and numbers on the cover page? Why does the language of the patent have to be so repetitious?

With these and other questions in mind, let us take a simple U.S. Patent and systematically break it down into all its constituent parts. We will examine each part in turn, discussing what purpose each part accomplishes, and then relate it to the entire document.

9.3 Dissecting a U.S. patent

Figure 9.1 is a reprint of the cover page of an issued patent, U.S. Patent No. 5,584,434. We will examine this patent in great detail, stopping to focus in on each significant feature. When we are done, you should have a very good understanding of how and why a patent is put together and where to find in it the exact information you need.

I selected this patent for three simple reasons. First, the invention it protects is so simple technologically that literally anyone can easily grasp it. Second, it is quite short (some patents are 50 or more pages long), which considerably simplifies our task. Nevertheless, you should find the same format in every patent, no matter how long or technologically complex. And the third reason? I am personally very familiar with this particular patent because I wrote the application and prosecuted it to issue before the Patent Office.

Incidentally, while this example is a U.S. patent, foreign patents and published patent applications follow a similar format. Once you are comfortable with the arrangement of a U.S. patent (or published application), you should be able to expand your patent-reading skills to foreign documents without a great deal of difficulty, provided, of course, they are printed in a language you can read. In fact, even if you do not understand Japanese or Russian, you still may be able to grasp something of the invention simply by paying close attention to the drawings and to the English abstract, if one is provided. Of course, if it is crucial that you understand a foreign-language patent thoroughly, then you may need to have it translated. However, the owners of many important inventions file for patents in a number of countries, and you may be able to find an English-language document equivalent to the one you cannot read. For more on how to find these gems, see Chapter 7.

9.4   The cover page

Do you notice the small numbers that appear in brackets in numerous places on this cover page? These numbers are in accordance with an international convention for labeling the various parts of almost any patent that is issued anywhere in the world, including the U.S.. We can use these numbers as a key to help us find and identify what we are looking at, even if the patent is printed in a foreign language.

In the patent of Figure 9.1, we see the number "45" that appears beside the name "Lipson." This tells us the last name of the inventor of this patent. This name is often used as an identifier, and we will follow that practice and refer to the patent from now on as "the Lipson patent." In this case, the patent names only one inventor. Supposing that it had, instead, multiple inventors, we would see "Lipson, et al."

Using the numerical labels mentioned above as our guide, we will now discuss each section of the cover page in turn, spending more time on those that are particularly informative:

- 19 tells us that this patent was granted by the U.S. Patent Office;
- 11 tells us that this patent is the 5,584,434th patent granted in the U.S. Patent Office;
- 45 tells us that the Lipson patent was issued on December 17, 1996. It will expire on January 25, 2015, 20 years after its filing date;
- 54 is the title of the patent — "DRINKING STRAW HAVING A CAGE FOR CONTAINING AN OBJECT THEREIN." This title is somewhat informative and gives us a rough idea of what the invention is about. But many patent titles (e.g., "ARTICLE HANDLING DEVICE") do not;
- 76 is the full name and address of the inventor;

United States Patent [19]

Lipson

[11] **Patent Number:** **5,584,434**

[45] **Date of Patent:** **Dec. 17, 1996**

[54] **DRINKING STRAW HAVING A CAGE FOR CONTAINING AN OBJECT THEREIN**

[76] Inventor: **Erik Lipson**, 1530 Locust St., #15F, Philadelphia, Pa. 19102

[21] Appl. No.: **377,851**

[22] Filed: **Jan. 25, 1995**

[51] Int. Cl.^{6} **A47G 21/18**
[52] U.S. Cl. **239/33; 446/74; 446/267**
[58] Field of Search 239/33, 24, 16; 446/267, 74; 215/229

[56] **References Cited**

U.S. PATENT DOCUMENTS

3,517,884	6/1970	Horvath	239/33
4,576,336	3/1986	Cohen	239/33

FOREIGN PATENT DOCUMENTS

3407733 9/1985 Germany .

Primary Examiner—Lesley D. Morris
Attorney, Agent, or Firm—Gifford, Krass, Groh, Sprinkle, Patmore, Anderson & Citkowski, P.C.

[57] **ABSTRACT**

A novelty drinking straw in the form of a drinking tube having a three dimensional cage for containing a prize object therein. The straw includes first and second linear end portions, a three dimensional, prize containing cage, and means for retaining the prize in the cage. The cage may be configured so that the prize cannot slip out, or it may include a linear extending portion passing through the cage and an apertured prize contained therein. The novelty straw and contained prize object find particular utility as giveaway premiums and favors.

12 Claims, 2 Drawing Sheets

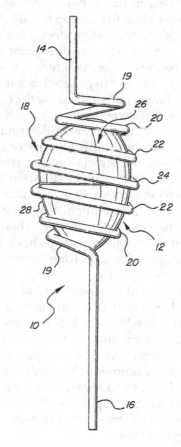

- 21 is the number assigned by the Patent Office to the application that was eventually issued as the Lipson patent while it was undergoing examination in the PTO. In this case, the application matured into a patent, but in many cases, Section 22 will contain information about predecessor applications that were abandoned by the inventor for any number of reasons or issued into other U.S. patents;
- 22 tells us when the Lipson patent application was filed;
- 23, which would give information about any foreign patent applications and their filing dates that impact the effective filing date and expiration date of the U.S. patent involved, does not appear on this patent. From this we may conclude that Lipson first filed for patent protection for his invention in the U.S.;
- 24, which also does not appear, would tell us the name of any assignee of the patent, in other words the name of the entity (typically a corporation, university, or government entity) that actually owns the patent. Bear in mind, however, that even if "Corporation X" is listed as the assignee, ownership could have been transferred to "Corporation Y" since the patent document was printed. We can find information concerning any subsequent transfer of ownership by searching the records of the Assignment Branch of the PTO (most conveniently done on-line at uspto.gov), but it will not appear on the face of an already-issued patent. Because the Lipson patent shows no assignee, we know that ownership rested in the inventor at the time the patent was issued, but we cannot determine the present ownership status simply from examining the cover page;
- 51 tells us the classes and subclasses into which the subject patent falls according to the technical classification scheme of The World Patent Organization in Geneva, Switzerland;
- 52 gives us the same information, this time using the system of classification of the U.S. Patent Office. This can be helpful if we want to find out what other patents have been issued in these subclasses that might have a bearing on the subject matter of the patent we are concerned with;
- 58 tells us the classes and subclasses of patents actually searched by the examiner in charge of the Lipson application;
- 56 lists the documents which the U.S. examiner found to be particularly relevant to determining whether the Lipson invention was entitled to patent protection. The U.S. patent documents are listed first, then the foreign documents. You will notice that the identifying number of each document appears first, then its date, then the last name of the inventor. Some patents also show the category "Non-patent References," and these are often journal articles, technical manuals, or sales literature;
- 57 is the Abstract of the invention and is the first place that we can learn considerable information about what the patent covers. It is a

concise summary of the most important aspects of the invention, typically expressed in broad terms;

- A single drawing appears on the cover page. The PTO determines which one of the drawing figures that appear in an application represents the best depiction of what the invention is all about. In a mechanical or device patent like the Lipson patent, it is often possible to get a lot of good information about what the patent covers just from looking at the cover page drawing;

- Separate the pages of a patent so that you can go back and forth between the Detailed Description and the drawing figures, finding where each numbered element of a figure is discussed in the text and vice versa.

9.5   The specification

The specification is the inventor's description of what he or she considers the invention to be. Most patent specifications include both a drawing component and a written component, although some patents (chiefly those in the chemical arts) omit drawings. The specification of the Lipson patent appears as Figure 9.2a through Figure 9.2d and Figure 9.3a through Figure 9.3c. Figure 9.2a through Figure 9.2d contain, respectively, drawings FIG. 1 through FIG. 4 of the Lipson patent, and Figure 9.3a through Figure 9.3c each correspond to one of the three pages of the written portion of the patent's specification. U.S. patents have been printed in a two-column-per-page format for many years, and two columns appear in each of Figure 9.3a through Figure 9.3c. You will also see a series of numbers running down the center of each page of the patent between the pair of columns. These numbers help you figure out which line of a column you are currently reading. When quoting language from a patent specification, the following format is used — "A prize object such as a zoo animal could be placed inside the 'bars' created by this embodiment" (col. 2, line 27 to line 28).

You will also notice at various places in Figure 9.3a through Figure 9.3c a series of headings in boldface type — **"FIELD OF THE INVENTION," "BACKGROUND OF THE INVENTION,"** etc. The headings, which are mandated by statute, serve as guideposts to help us navigate our way through any U.S. patent.

In the following sections we will take a brief tour of the specification of the Lipson patent.

9.5.1   The drawings

Looking first at Figure 9.2a through Figure 9.2d, the four drawings of the Lipson patent, we immediately notice that Figure 9.2a is familiar to us because it is the same one the PTO selected to appear on the face of the

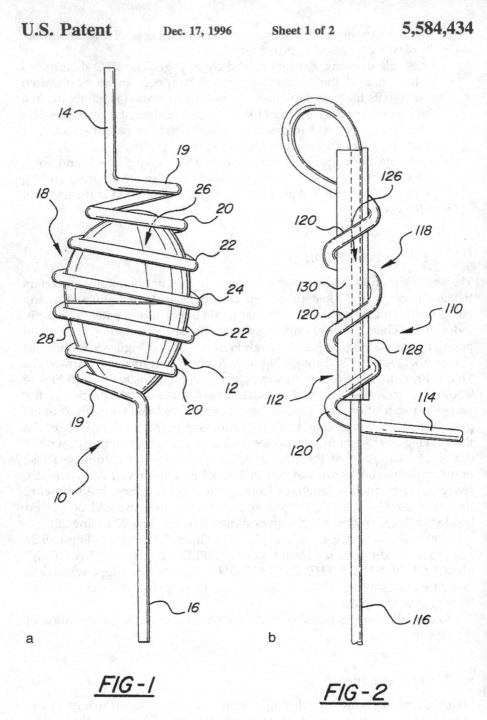

Figure 9.2a through Figure 9.2d   Reprint of U.S. Patent 5,584,434 FIG. 1 through FIG. 4, respectively.

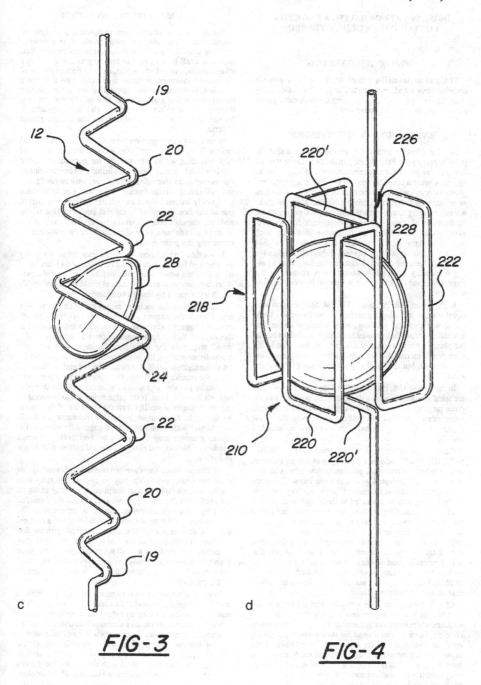

FIG-3

FIG-4

Figure 9.2   (continued)

5,584,434

| 1 | 2 |

DRINKING STRAW HAVING A CAGE FOR CONTAINING AN OBJECT THEREIN

SUMMARY OF THE INVENTION

The present invention has been designed to overcome the deficiencies in the prior art noted above. It is a novelty drinking straw for use in combination with a prize object, such as a small ball, a two-part hollow plastic egg having a token therein, a rolled up coupon, etc. The drinking straw of the present invention includes a drinking tube defining a hollow, continuous flow passage and having first and second linearly extending end portions. One of the end portions defines a mouthpiece. Preferably, the drinking tube is formed of a resilient, translucent or transparent plastic material.

FIELD OF THE INVENTION

The present invention relates to the field of novelty drinking straws and, more particularly, to such a drinking straw having a portion defining a three dimensional cage for containment of a novelty prize therein.

BACKGROUND OF THE INVENTION

Novelty drinking straws are becoming more and more popular every year. Part of their popularity is due to the fact that they are inexpensive to manufacture and easy to package, thus being a desirable product to serve as a "premium" given away at such establishments as fast food restaurants, or to serve as favors at children's parties. Many of these novelty drinking straws include drinking passages which are in the form of loops, spirals, flowers, stick figures, eyeglasses, etc. Some examples of patents disclosing such novelty drinking straws include U.S. Pat. Nos. 3,517,884 and 4,576,336, as well as German Published Application 3407733. These references all disclose drinking straws having a plurality of spirally wound loops connecting the mouthpiece and the end which is inserted into the liquid to be drunk.

In particular, U.S. Pat. No. 3,517,894 discloses one such novelty straw including a plurality of spirally wound loops connecting two end portions, each having a constricted internal diameter. One or more small objects, such as beads, are placed within the coils of the spiral portion. Due to the constricted internal diameters of the ends, fluid may pass therethrough, but the small objects are retained inside the straw.

In the case of U.S. Pat. No. 4,576,336, a plurality of coils are spirally wound around a drinking glass, with the mouthpiece projecting from outside the glass, and the other end being submerged inside the glass. In this way, the straw engages the glass so the two can be picked up simultaneously.

Furthermore, applicant in the invention of U.S. Pat. No. 5,427,315, which discloses a drinking straw having an insert and including a continuous passage hollow tube with two ends which is interrupted by one or more loops. The loops are oriented parallel to the ends of the passageway. One end of the passageway is connected to the looped portion by a transversely extending portion. The straw also includes a planar insert having an aperture or notch for engagement with the transverse portion so that the insert may be attached to the straw. However, due to the arrangement and orientation of the spirals, the novelty drinking straw of the disclosed patent is most advantageously used in combination with planar configured novelty items (printed pasteboard, cardboard or plastic), thus somewhat restricting its usefulness as an entertainment device.

Clearly, there is a need for a novelty drinking straw which may be used in combination with a variety of toys and other amusement devices so as to create an item of great appeal to children. There is also a need for such a novelty drinking straw in combination with a retained object wherein the object retained therein remains intact, and easily removable from the straw. Finally, there is a need for such a drinking straw which is inexpensive and easy enough to manufacture so that it may be used as a free, promotional item.

In an intermediate portion between said first and second end portions, the drinking tube defines a three dimensional cage including an interior volume for containing a prize object which is dimensioned to fit inside the interior volume. For example, the three dimensional cage may be configured as a cylindrical volume formed by a plurality of coils which are spirally wound around an axis defined by or parallel to one of the first or second linear end portions. That is, the coils are disposed roughly transverse the linear axis of the straw so as to define a three-dimensional interior volume for containing the prize object.

In another embodiment, the cage may define a roughly parallelepiped interior volume and be formed of a plurality of linear portions which wind back and forth and up and down. A prize object such as a zoo animal could be placed inside the "bars" created by this embodiment.

The invention also includes means for retaining the prize object inside the interior volume. In one preferred embodiment, the cage is comprised of loops which are graduated in the sizes of their circumferences, from a largest loop disposed proximate the middle of the cage, with the coils gradually decreasing in circumferential size towards smallest circumference coils displaced at either end of the cage. Thus, the interior volume has its largest diameter proximate its middle portion, and gradually tapers down in diameter towards its ends. Thus, if the prize object is dimensioned so that its diameter is smaller than the diameter of the middle portion of the cage, but larger than the end diameters, it will be retained within the egg shaped cage. Alternatively, the retaining means may be formed by horizontally oriented windings which extend along the top and bottom of the cage to enclose it.

Of course, since the cage is composed of windings or coils, the windings or coils may be displaced linearly with respect to each other, so that the prize object may easily be removed from the cage by passing it between the displaced windings. This removability feature is enhanced if the novelty straw of the present invention is formed of a resilient material, such as plastic. The user may simply grip the first and second ends and pull them in opposite directions, thus spreading the windings and allowing removal of the prize object, or, alternatively, may bend the two ends toward each other so as to fan out the windings, thus enabling retrieval of the prize object.

In another preferred embodiment, the cage comprises a plurality of spirally wound coils which are approximately of equal size to define a roughly cylindrical interior volume. In this embodiment, the means for retaining an object inside the cylindrical interior volume comprises a linear extension of one of the two end portions which extends through the helical cage, itself and joins the cage to the one end portion. Typically, a prize object used with this embodiment has a through passage so that the linear extension of the drinking tube will pass through the object, itself, thus helping to retain

Figure 9.3a through Figure 9.3c   Reprint of U.S. Patent 5,584,434, text column 1 and column 2 (Figure 9.3a), text column 3 and column 4 (Figure 9.3b), and text column 5 and column 6 (Figure 9.3c).

5,584,434

3

it within the cage. Retention of such an object is assisted by the resiliency of the drinking tube which causes the coils to grip the object, thus preventing it from slipping down the tube. This embodiment of the present invention is particularly useful if the prize object is a rolled up piece of paper or cardboard having a coupon, picture, game, or other message printed thereon. The piece of paper or cardboard is simply rolled into a tube, slipped over one end of the straw, and slid into the helical cage.

BRIEF DESCRIPTION OF THE DRAWINGS

The following detailed description is best understood with reference to the following drawings in which:

FIG. 1 is a perspective view of one embodiment of the novelty drinking straw of the present invention;

FIG. 2 is a perspective view of a second embodiment of the drinking straw of the present invention;

FIG. 3 is a partial perspective view of the embodiment of FIG. 1 showing the coils spread apart to enable retrieval of the prize object; and

FIG. 4 is a perspective view of yet another embodiment of the drinking straw of the present invention showing a parallepiped cage.

DETAILED DESCRIPTION OF THE PREFERRED EMBODIMENTS

Throughout the following detailed description, like numerals are used to reference the same element of the present invention shown in multiple figures thereof. Referring now to the drawings, and in particular to FIGS. 1 and 3, there is shown a novelty drinking straw 10 including a drinking tube 12 defining a hollow, continuous flow passage therethrough. The tube includes first 14 and second 16 linearly extending end portions. In the depicted embodiment, it is contemplated that the first linear end portion 14 will be used as a mouthpiece and second end portion 16 will be submerged in the liquid to be drunk.

A helical cage 18 is disposed intermediate the first and second end portions 14, 16. It includes a plurality of spirally wound coils 19, 20, 22, 24. As can readily be seen by examining FIG. 1, the coils gradually increase in size from either end coil 19 toward the middle coil 24. Thus, the coils 19–24 collectively define an interior volume 26 which is broader in its middle than it is at either end.

An egg shaped prize object 28 is shown in place inside helical coil 18. It has a circumference in at least one direction that is greater than the internal circumference of end coils 19 but less than the internal circumference of middle coil 24. Thus, due to the described configuration of interior volume 26, prize object 28 is retained therein.

It is contemplated that the embodiment of the present invention depicted in FIG. 1 may be given away as a prize or promotional item, or sold for a small sum. To that end, its desirability as a prize is enhanced if the prize object 28 is of the familiar hollow, two part form so that it may contain yet another toy or surprise therein. Thus, upon receiving the novelty straw of the present invention, a child may enjoy it in several ways: the drinking tube 12 may be used by itself or with the prize object 28 still retained therein, and the child may enjoy watching passage of liquid through the coils 19–24. Also, the child may remove the prize object 28 from the helical cage. The child may then open the "egg" to discover the surprise therein.

4

FIG. 3 depicts one way in which the cage 18 may be opened so that the prize object 28 can be removed. In this case, the two end portions 14, 16 have been linearly displaced in opposite directions with respect to each other so as to spread apart the coils 19–24. In this way, the prize object 28 may be easily removed from the cage 18 by passing it between the spread apart coils. Another way (not depicted) would be to bend the straw so that the two end portions 14, 16 lie adjacent each other, thus causing the coils 19–24 to fan out and permit removal of the prize object 28.

Of course, a different kind of prize object than the hollow egg depicted may be placed inside the helical cage 18. For example, a small toy ball may be placed therein, a plastic figurine or action figure, a toy vehicle, etc. Thus, the straw of the present invention is not limited to the depicted straw/prize object combination, but may be used to contain a staggering variety of toys and other items particularly appealing to children.

An alternative embodiment 110 of the straw of the present invention is depicted in FIG. 2. Like the previously described embodiment, straw 110 also includes a drinking tube 112 which defines a hollow, continuous flow passage. It further includes first 114 and second 116 end linear portions, although it should be noted that, in this embodiment, the two end portions 114, 116 are oriented perpendicular, rather than parallel, to each other. The embodiment of FIG. 2 also includes a helical cage 118 comprised of a plurality of spirally wound coils 120 which are approximately equal sized, thus defining a roughly cylindrical interior volume 126.

Extending linearly from second end portion 116 is a linear extension 130 (shown in phantom). It connects second end portion 116 with helical cage 118. Furthermore, it passes through prize object 128 which, in the depicted embodiment, is in the form of a rolled up piece of paper, with linear extension 130 passing through the hollow middle of the paper roll 128.

Yet another embodiment 210 of the novelty drinking straw is shown in FIG. 4. In this embodiment, the three-dimensional cage 218 defines a roughly parallepiped interior volume 226. The cage 218 is formed of a plurality of horizontally and vertically oriented windings 220,222, respectively, which together define the flat sides of the cage 218. Two of the horizontal windings 220' are disposed on the top and bottom of the cage 218 to enclose volume 226 and retain a prize object in the form of a ball 228 therein.

Preferably, the novelty straw of the present invention is fabricated from a resilient plastic, such as polyethylene tubing. Thus, the resilient coils 120 will grip the paper roll 128 to assist in securing it in the straw 110. In this embodiment, the object may be removed from the straw by simply slipping the paper roll 128 down the extended portion 130 and off second end portion 116.

Again, the embodiment depicted in FIG. 2 is not limited to a combination novelty straw and paper roll; the prize object may take a number of other forms as long as it includes a through passage so that the extended portion 130 may be passed thereto to secure the prize object to the straw.

Thus, a novelty drinking straw including a drinking tube having a three dimensional cage to enclose a prize object has been depicted and described with reference to certain embodiments thereof. Doubtless, one skilled in the art, having had the benefit of the teachings of the present invention, may configure the drinking tube somewhat differently and may use the straw with prize objects differing from those depicted without departing from the scope of the

Figure 9.3b

5,584,434

5

6

present invention. For example, instead of defining a cylindrical, egg shaped or parallepiped volume, the three dimensional cage of the straw may define volumes of other and more complex configurations. Thus, it is the claims appended hereto, and all reasonable equivalents thereof, rather than the depicted embodiments and exemplifications, which define the true scope of the present invention.

I claim:

1. A novelty drinking straw comprising:

a drinking tube defining a hollow, continuous flow passage and including first and second linearly extending end portions, one of said first and second end portions defining a mouthpiece;

a three dimensional cage formed by said flow passage and disposed therealong intermediate said first and second end portions, said cage defining an interior volume configured to retain a prize object entirely therein; and

means formed by said passageway for retaining said object within said interior volume.

2. The drinking straw of claim 1 wherein one of said first and second end portions lies parallel to an axis and said cage is formed of a plurality of coils wound spirally around said axis.

3. The straw of claim 2 wherein said plurality of coils each has a circumference, said coils being graduated in circumference from a largest circumference coil proximate a middle portion of said cage to a smallest circumference coil disposed proximate each end of said cage to form said means for retaining said object therein.

4. The straw of claim 1 wherein said cage is formed of a plurality of horizontally and vertically extending windings to define a parallepiped interior volume.

5. The straw of claim 1 wherein said drinking tube is formed of a resilient material.

6. A combination novelty drinking straw and prize object contained therein, said combination comprising:

a resilient drinking tube defining a hollow, continuous flow passage and including first and second linearly extending end portions, one of said first and second end portions defining a mouthpiece

a helical cage having first and second ends and a middle portion medial thereof, said cage being formed by said flow passage and disposed therealong intermediate said first and second end portions, said cage defining an interior volume and being formed by a plurality of coils wound spirally around said volume, said plurality of coils each having a circumference and being graduated in circumference from a largest circumference coil proximate said middle portion of said cage to a smallest circumference coil disposed proximate each of said first and second end of said cage; and

a novelty prize object disposed in said interior volume, said prize being dimensioned so as to be normally retained within said interior volume and to be removable from said interior volume when adjacent pairs of coils of said cage are linearly displaced away from each other so that said object prize is passable therebetween.

7. The straw of claim 6 wherein the object has a circumference in at least one direction sized smaller than said largest coil circumference and larger than said smallest coil circumferences.

8. The straw of claim 6 wherein said drinking tube is formed of a resilient material.

9. A combination novelty drinking straw and prize object contained therein, said combination comprising:

a drinking tube defining a hollow, continuous flow passage and including first and second linearly extending end portions, one of said first and second end portions defining a mouthpiece and lying parallel an axis;

a three dimensional cage having first and second ends and a middle portion medial thereof, said cage being formed by said flow passage and disposed therealong intermediate said first and second end portions, said cage defining an interior volume for containing a prize object dimensioned to fit therein, said cage being formed of a plurality of coils wound spirally around said axis, said coils being graduated in circumference from a largest circumference coil proximate said middle portion of said cage to a smallest circumference coil disposed proximate each of said first and second ends of said cage to form means for retaining said object therein, wherein the prize object has a circumference in at least one direction sized smaller than said largest coil circumference and larger than said smallest coil circumference.

10. A novelty drinking straw comprising:

a drinking tube defining a hollow, continuous flow passage and including first and second linearly extending end portions, one of said first and second end portions defining a mouthpiece;

a three dimensional cage formed by said flow passage and disposed therealong intermediate said first and second end portions, said cage being formed of a plurality of horizontally and vertically extending windings to define a parallepiped interior volume or containing a prize object dimensioned to fit therein; and

means formed by said passageway or retaining said object within said interior volume.

11. A combination novelty drinking straw and prize object contained therein, said combination comprising:

a drinking tube defining a hollow, continuous flow passage and including first and second linearly extending end portions, one of said first and second end portions defining a mouthpiece;

a three dimensional cage formed by said flow passage and disposed therealong intermediate said first and second end portions, said cage defining an interior volume for containing a prize object having means defining a passage formed therethrough and dimensioned to fit in said cage; and

means formed by said passageway tier retaining said object within said interior volume comprising a linear extension of said one end portion which extends along said axis through said helical cage for extension through said passage of said prize object when said prize object is placed within said interior volume.

12. The straw of claim 11 wherein said coils are substantially uniform in size.

*   *   *   *   *

Figure 9.3c

patent. It is a good idea to pay particular attention to the drawing chosen for the cover because it generally contains the clearest depiction of the broad concept of the invention. Thus, in Figure 9.2a we find a drawing of what looks like a hollow tube or drinking straw that is both like and unlike straws with which we are already familiar. It appears to have an upper sipping end and a second end to be placed in the liquid. What is unusual and unfamiliar about this particular straw is that a series of elliptical winding is disposed between the two ends. The shape formed by this winding is roughly ovoid, and we se some kind of egg-shaped object contained in the coils. We can readily understand simply from studying this one drawing something important about the invention of the Lipson patent. We already know from the title that the invention is a drinking straw and that it includes a "cage for containing a prize object." From Figure 9.2a, we now understand that a portion of the straw forms the cage, itself. In other words, the prize-containing cage is not a structure formed separately from the straw and later attached to it, but is integral with it.

We also cannot help but notice that the straws depicted in the other three drawings are all quite different from the straw of Figure 9.2a. For example, in Figure 9.2b we see that a portion of the straw extends through the prize object (in this case, what appears to be a rolled-up piece of paper), and the coils wind around it. In Figure 9.3d, the straw is wound around in such a way so as to form a cube-shaped cage. These two drawings help us understand that the invention is probably broader than simply the straw and ovoid cage depicted in Figure 9.3a, but may include a variety of other shapes. We also may be somewhat puzzled by the appearance of Figure 9.2d because it does not seem that the straw windings in that configuration could actually hold or contain the prize object. But we must always keep in mind that the drawings are meant to be read in conjunction with the Detailed Description, so we will put that drawing aside for later consideration.

You will see drawings somewhat like these in all U.S. and foreign patents that cover devices and components (whether mechanical, electrical, or medical), structures, systems comprised of various subcomponents, and the like. In contrast, patents for chemical compositions typically do not contain drawing figures, although formulas and molecular structures may appear in the body of the patent.

When present, the drawings should be completely mined to reveal every bit of helpful information they contain. It is often the case that the drawings offer a clearer depiction of the invention than the written description, particularly if the document is a translation of a patent originally written in a foreign language and first filed abroad. Whatever you do, do not pass over them lightly.

You will notice that each drawing contains several numerical labels that reference various elements of the invention in each figure. For consistency's sake, the PTO requires that the same number be associated with the same element through all the figures. Moreover, you will also find these same numerical labels sprinkled throughout the Detailed Description section of the written text. It is a good idea to separate the pages of a patent so that you can go back and forth between the Detailed Description and the drawings, finding where each element of a figure is discussed in the text and vice versa.

Certain conventions are used in drafting the figures. Most labels include a leader line. If the label free end touches some part of the figure, it means the label refers only to that specific part. For example, reference numeral 14 in the Lipson patent drawings refers specifically to the upper end of the straw. If the leader line terminates in an arrowhead, that means the label refers generally to the entire structure toward which the arrowhead points. Thus, reference numeral 10 refers to the entire object, including the hollow tube, the three-dimensional cage, and the prize object contained therein. If the label is an underlined number, you will usually find it on top of the element it references. Schematic drawings of the type used in flowcharts and electrical circuits often employ this kind of labeling. Broken lines that appear in the drawings (e.g., Figure 9.2b) show hidden structures — ones that would not normally be visible from that perspective. Furthermore, cross-sectional views, part cross sections, detail views, graphs, charts, printouts, DNA segments, actual photographs, and many other types of pictorial depictions may appear as figures in patent documents. Some skill in blueprint reading is often helpful in deciphering particularly complex drawings. If you do not happen to have that skill, you might want to ask one of your coworkers for assistance.

Moving on to Figure 3a through Figure 3c, the written portion of the Lipson patent specification, we encounter the following first heading.

9.5.2   Field of the invention

This section tells us the broad field of technology (novelty drinking straws) into which the invention falls, and also a little bit about what the invention is (a straw that includes a three-dimensional cage for containing a novelty prize).

Next we find the second heading.

9.5.3   Background of the invention

This section usually contains a fairly brief discussion of some prior art the inventor considers to have a bearing on his or her own invention, either because it describes something somewhat similar, because it represents a different way to solve the problems the inventor had in mind when making the invention, or because it describes one particular feature of the invention, though used in a different context. Again, the prior art discussed in a typical

Background may be either patent or nonpatent documents. Frequently the prior art includes patents granted to the same inventor describing an earlier version of a similar inventive concept.

For example, in the third full paragraph of Lipson's Background section (Figure 9.3a), we find a discussion of an issued U.S. patent (4,576,336) showing a novelty straw similar to that depicted in Figure 9.1 of the Lipson patent in that the straw includes a plurality of coils. Instead of forming a three-dimensional cage into which a prize object may be put, the coils of the older patent are wrapped around a drinking glass so that the straw and glass may be picked up together. Lipson considers this to be structurally different from his invention and useful for a different purpose. The examiner in charge of the Lipson application must have agreed because the Lipson patent was allowed to be issued despite the existence of this prior patent. You cannot, of course, get a new patent for something that has already been invented, or even for something that is so similar to the prior art that it would have been obvious enough for someone of skill in that particular technology to think of it.

Pay particular attention to the last paragraph or two of the Background section of any patent. In it you will usually find a short discussion of what the prior art discussed earlier fails to achieve and what problem the present inventor had in mind when the invention was made. This can be very helpful to you in coming to grips with the "kernel" or "gist" of the inventive concept disclosed in the patent.

9.5.4 *Summary of the invention*

A properly written invention Summary is structured somewhat like a newspaper story. That is, the most important information is contained in the first paragraph or two, and each subsequent paragraph reveals the invention in finer and finer detail. However, it is critically important to remember that it is *not* the Summary section that actually defines the invention protected by the patent. That is the function of the Claims, which we will discuss in greater detail below.

More often than not, the invention described in the Summary section — and this is especially true of the first few paragraphs — is broader than the claimed invention, and sometimes so much broader that it would be hard to even guess the breadth of the claims that are finally issued. For this reason, we should always approach the Summary with some caution, looking to it for a general idea of what the inventor *thought* was his or her invention when the patent application was filed, but always realizing that the invention actually defined by the claims has probably become somewhat narrower by the time the application has made its way though the examination process in the Patent Office.

In the case of the Lipson patent, we see from the first several paragraphs of the Summary that the inventor had a very broad inventive concept in mind when he filed his application. We read in the first paragraph that the invention is a straw for use in combination with an object. The "prize object," which is not necessarily included as part of the invention, is broadly defined and may include such items as "a small ball, a two-part hollow egg, a plastic egg having a token therein [the Figure 9.2a version], a rolled-up coupon [the Figure 9.2b version], etc." (col. 2, line 5 to line 6).

In the next paragraph, we see that the straw, in addition to having the sipping end and the liquid immersion end typical of all drinking straws, also has an additional feature — a three-dimensional cage for containing the prize object. We can tell that the cage does not have to be any particular shape because at least two different ones are explicitly described — a cylinder and a parallelepiped.

In the fourth paragraph, we read that there is yet another element to the invention — means for retaining the prize within the cage. The next two paragraphs describe two very different ways this can be accomplished. The cage may be comprised of spiral windings of the straw that taper down from a wide middle portion to two narrower ends. These ends trap the prize object inside the resultant ovoid cage. Alternately, a portion of the straw, itself, may extend through a portion of the prize object, thus securing the prize within the cage.

After we have read through the fairly short Summary section of the Lipson patent, we now have a pretty good idea of the nature of the invention — the elements that combine together to make its structure and how this structure functions to accomplish the job of retaining the prize object in association with the straw, as well as various embodiments, or versions, of the invention that differ considerably in appearance and shape. However, this may not be the case with more technically complex patents where the Summary often is too simplified to be of great value.

The technology involved in the Lipson patent is "childishly" simple. It is relatively easy for even a patent novice to read and understand the patent specification, and to grasp the nature of the invention rather quickly. However, most patents describe inventions that are far more complex. Mechanical patents may disclose devices having a hundred or more parts. Chemical patents may include laundry lists of possible components. The technology discussed in biotech or electronic systems patent may be mind-numbingly complex. The more complex the invention and the technology behind it, the more likely it is that you will have to methodically consider each area of the patent document, scrutinizing the drawings and written description for the clues you will need to understand the invention as a whole.

9.5.5 Brief description of the drawings

This section should be read in conjunction with the drawings we have
already discussed. Basically, it is a series of short paragraphs, each briefly
describing one of the drawings. These short descriptions can help us understand otherwise puzzling drawings. For example, we are told that FIG. 3
(our Figure 9.2c) "is a partial perspective view of the embodiment of FIG. 1
showing the coils spread apart to enable retrieval of the prize object."
Looking back at this drawing with that additional information in mind,
we now can understand much more clearly what aspect of the invention
this figure is intended to depict — the ready removability of the prize object
from the cage by simply stretching out the coils of the straw.

As mentioned before, many patent drawings are complex and include
partial cross sections, partial detail views, exploded views, etc. Without
looking at the corresponding paragraph of the Brief Description, we may
have difficulty understanding drawings of more complex nature, and their
role in illustrating some important aspect of the invention may completely
escape us. So the Brief Description should be regarded as yet another tool
that brings us a better understanding of what invention the patent covers.

9.5.6 Detailed description of the preferred embodiments

Unlike the Summary section, the Detailed Description is written to correlate
with the drawings. Typically, the first or second paragraph of this section
will say something like "Referring now to the drawings and in particular to
Figs. 1 and 2, here is shown the widget 10 of my invention, including a
hollow widget body 12 and a pair of opposed fins 14 disposed at a first end
16 of the body 12." Generally, the features of the drawings will be described
in more or less numerical sequence, with the first embodiment of the invention that appears in the Detailed Description almost always bearing the
reference numeral "10." Subsequent embodiments, when present, are often
referenced by numerals "110," "210," and so forth. In fact, we find this to be
the case with the Lipson patent. The first embodiment, which is shown in
FIG. 1 and FIG. 3 (our Figure 9.2a and Figure 9.2c) is referenced by numeral
"10," the second embodiment shown in FIG. 2 (our Figure 9.2b) by numeral
"110," and the third and final embodiment shown in FIG. 4 (our Figure 9.2d)
by numeral "210." You will also notice that the remaining elements of each
embodiment follow a similar pattern. For example, the first embodiment, the
drinking tube of FIG. 1 and FIG. 3, is referenced by numeral "12," the tube of
FIG. 2 by numeral "112," and the tube of FIG. 4 by numeral "212."

Occasionally, you will see drawings labeled "Prior Art." They are used
to help illustrate some critical difference between typical prior art devices
and the improvement protected by the patent. Such drawings usually employ the convention of labeling the elements with single digit reference
numerals so that numeral "10" will remain available to reference the first

embodiment of the actual invention. Usually you will find the discussion of any prior art drawings that may be present appearing in the earliest paragraphs of the Detailed Description.

As mentioned earlier, the pages of the patent should be separated so that you can easily compare each drawing with its matching description as you proceed, paragraph by paragraph, though the Detailed Description. Reading this section of a patent, which can be quite lengthy, may seem tedious to you and that may often be the case. Fortunately, you will find that you often can acquire a good enough understanding of the invention from carefully reading the Abstract, broadest claims, and Summary, as well as from examining the drawings, so that you will be able to move through the Detailed Description very quickly. How much attention you need to pay to this section will largely depend on the problem the particular patent poses for you, as well as on how clearly the patent is written.

For example, if you are an inventor and you want to know if some earlier inventor "got there first," you should read as much of the patent as you need to figure out whether what it describes is the same as (or so similar to) your invention so as to make it unlikely that you will be able to patent your invention. Many, if not most, of the time, you will be able to make this determination early on. The drawings, alone, may tell you everything you need to know. On the other hand, you might be more interested in whether you can safely put your new product on the market without concern about infringing someone else's patent. That situation requires you to dope out what the claims of the patent actually cover. And the claims may cover an invention considerably narrower than you would expect from simply examining the drawings and skimming through the Summary. When it comes to questions of claim coverage and claim interpretation, you will find yourself relying heavily on a thorough reading of the Detailed Description.

A special note to those who are reading a patent primarily for the technical information it contains: since the detailed description provides the real "meat" of the disclosure, it is highly likely you will be concentrating most of your efforts in this section. While patents can be very valuable as teaching documents (that is why the government grants inventors a limited monopoly to practice their invention; in return the public learns the details of how to practice the invention, thus advancing the state of knowledge in that particular technical field), you should keep in mind that a patent is not as reliable as, say, a peer-reviewed journal article on the same subject. There is a certain amount of technical creativity that goes into a lot of patent applications, and some of the explanations of how and why the invention works are more speculative and theoretical than factual. Some healthy skepticism is useful when approaching a patent disclosure primarily to learn what it teaches.

In addition to a complete description of each element in each drawing, the Detailed Description will usually explain how the devices depicted in the drawings actually operate, for what purposes they may be used, and the advantages their use offers over the prior art. Also, you are likely to see brief mentions of alternative structures (screws instead of nails, hook-and-loop fastener instead of glue, etc.) that could be used in the invention with similar effect, or different configurations in which some of the structures of the invention could be shaped without changing the basic nature of the invention. It would simply be impractical to describe all conceivable variations of this sort in detail and still keep the patent document to a reasonable length. Yet, an inventor (who is obligated by law to describe the best way of practicing the invention in a complete enough way to enable someone "of skill in the art" to practice it) needs to make it clear that the patent covers something broader than exactly what the drawings show. So a patent will usually make only brief reference to these sorts of design variations that are too trivial to be fully described.

At the end of the Detailed Description section of the Lipson patent, we find another convention typically used to try to ensure that the patent description is not unduly narrow:

> Doubtless, one skilled in the art, having had the benefit of the teachings of the present invention, may configure the drinking tube somewhat differently and may use the straw with prize objects differing from those depicted without departing from the scope of the present invention. For example, instead of defining a cylindrical, egg shaped or parallepiped volume, the three-dimensional cage of the straw may define volumes of other and more complex configurations. Thus, it is the claims appended hereto, and all reasonable equivalents thereof, rather than the depicted embodiments and exemplifications, which define the true scope of the present invention.

You will find language similar to this at the end of the Detailed Description section of almost all U.S. patents. To put it in much simpler words, the scope of the patent is not limited to what is shown in the drawings and described in the specification. To understand what technology a patent protects, you must always look to its claims.

9.5.7   The claims

Turning finally to the claims of the Lipson patent, you will notice that they number twelve in total, with claim 1, claim 6, and claim 9 through claim 11 being what are called *independent* claims. You can identify them as independent claims because they are complete in themselves and do not refer to any other claims. On the other hand, claim 2 through –claim 5, claim 7, claim 8, and claim 12 are *dependent* claims. Each of them specifically refers to a previous claim. For example, claim 2 refers to claim 1 ("The drinking straw of claim 1..."), and claim 3 refers to claim 2 ("The straw of claim 2 ... ''). Another way of saying this is that claim 2 depends on claim 1, while claim 3 depends on both claim 2 *and* claim 1 (because it depends on claim 2, which, in turn, depends on claim 1). Continuing through the Lipson claims, you see that dependent claims 4 and 5 depend on independent claim 1, dependent claims 7 and 8 depend on independent claim 6, and, finally, dependent claim 12 depends on independent claim 11. Independent claims 9 and 10 stand alone — no other claims depend on them.

What does it mean for a claim to depend on another claim? It is just a shorthand way of saying that the dependent claim includes all of the features listed in the claim on which it depends. For example, the Lipson claim 5 includes all of the features of claim 1 (the drinking tube, the three-dimensional cage, and the retaining means as described in claim 1), but also has the additional feature that the drinking tube must be formed of a resilient material. If a dependent claim depends on another dependent claim (such as claim 3 on claim 2), then it automatically includes all of the features of all the dependent claims on which it depends, as well as all the features of any claims these claims ultimately depend on. It is not unusual in patents to find strings of claims wherein a single dependent claim may incorporate the features of four, five, six, or even more previous claims.

While all this dependent claim business may seem confusing, it need not trouble you all that much when you are attempting to read a patent. The independent claims of a patent are far more important because each of them is broader in terms of patent protection than any of the claims that depend on them. Thus, claim 2 through claim 5 are all narrower than claim 1 because they each include an additional limitation to the invention. If we want to understand what technology a patent covers, then we must always look first to the independent claims, and we should first try to find the broadest and the most representative of the independent claims.

Let us look at the five independent claims of the Lipson patent with an eye to deciding which is the broadest and which is the most representative. What factors can help us make these determinations?

The first factor to look at is the numbering of the claims. Very often, claim 1 will best represent the invention the inventor had in mind when the patent application was drafted. It is usually the first claim written by the drafter. Often, as in this patent, it will have a significant string of claims dependent on it. In cases where there are several depicted embodiments, claim 1 is most likely to cover the first depicted embodiment. Thus, you find that the Lipson claim 1 does, indeed, cover the embodiment depicted in FIG. 1 (our Figure 9.2a). Claim 1 requires three separate structures:

1. A novelty drinking straw comprising:
    a drinking tube defining a hollow, continuous flow passage
    and including first and second linearly extending end por-

tions, one of said first and second end portions defining a mouthpiece;

a three-dimensional cage formed by said flow passage and disposed therealong intermediate said first and second end portions, said cage defining an interior volume configured to retain a prize object entirely therein; and

means formed by said passageway for retaining said object within said interior volume.

Since all the depicted embodiments include a hollow drinking tube with extending end portions, one of which is a sipping end, the first clause of claim 1 reads on all the straws shown in the patent. The same can be said of the retaining means of the claim's last clause. In the case of FIG. 1 and FIG. 4 embodiments (our Figure 9.2a and Figure 9.2d), the retaining means is formed by the way the three-dimensional cage is shaped, either egg-shaped or cube-shaped, so that the prize object is confined therein. However, in the case of FIG. 2 embodiment (our Figure 9.2b), the retaining means is actually comprised of a turning of the straw that passed through the center of the paper roll, thereby trapping it inside. Nevertheless, the last clause of claim 1 is broad enough to cover all these embodiments.

But let us look more carefully at the middle clause of claim 1, the one that describes the three-dimensional cage:

a three-dimensional cage formed by said flow passage and disposed therealong intermediate said first and second end portions, *said cage defining an interior volume configured to retain a prize object entirely therein*; and

If we ignore the portion in italics, we can agree that the rest of this clause covers all three depicted embodiments. But, of course, we cannot ignore it because every word of a claim must be given due consideration in our attempt to decipher the claim. The italicized phrase requires that the prize object be entirely contained within its interior volume. A glance at FIG. 2 shows this is not the case for this embodiment. From that we must conclude that while claim 1 may be the most representative claim in the patent, it is not broad enough to cover all the depicted embodiments. Can we find a broader one?

Let us go down to the next independent claim, claim 6:

6. A combination novelty drinking straw and prize object contained therein, said combination comprising:

a resilient drinking tube defining a hollow, continuous flow passage and including first and second linearly extending end portions, one of said first and second end portions defining a mouthpiece;

a helical cage having first and second ends and a middle
portion medial thereof, said cage being formed by said flow
passage and disposed therealong intermediate said first and
second end portions, said cage defining an interior volume
and being formed by a plurality of coils wound spirally
around said volume, said plurality of coils each having a
circumference and being graduated in circumference from a
largest circumference coil proximate said middle portion of
said cage to a smallest circumference coil disposed proximate
each of said first and second end of said cage; and

a novelty prize object disposed in said interior volume, said
prize being dimensioned so as to be normally retained within
said interior volume and to be removable from said interior
volume when adjacent pairs of coils of said cage are linearly
displaced away from each other so that said object prize is
passable therebetween.

You will immediately notice that this claim, unlike claim 1, has four instead
of three elements and includes the prize object itself. Moreover, it is quite a
bit longer than claim 1, a good indication that it is likely to be a narrower,
not a broader, claim. Indeed, if we examine its second clause, we find
language in there that requires the three-dimensional cage to be formed of
helical spirals graduated in size. While this claim seems to cover the FIG. 1
embodiment, it clearly leaves out of its compass both other depicted cage
shapes.

Well then, what about claim 9, the next independent claim?

9. A combination novelty drinking straw and prize object con-
    tained therein, said combination comprising:

    a drinking tube defining a hollow, continuous flow passage
    and including first and second linearly extending end por-
    tions, one of said first and second end portions defining a
    mouthpiece and lying parallel an axis;

    a three-dimensional cage having first and second ends and a
    middle portion medial thereof, said cage being formed by
    said flow passage and disposed therealong intermediate
    said first and second end portions, said cage defining an
    interior volume for containing a prize object dimensioned to
    fit therein, said cage being formed of a plurality of coils
    wound spirally around said axis, said coils being graduated
    in circumference from a largest circumference coil proximate
    said middle portion of said cage to a smallest circumference
    coil disposed proximate each of said first and second ends of
    said cage to form means for retaining said object therein,

wherein the prize object has a circumference in at least one direction sized smaller than said largest coil circumference and larger than said smallest coil circumference.

This time, we find no mention of a prize retaining means, so we might guess this claim is broader than claim 1 and claim 6, but we would be wrong. You will note that the second clause is very long and contains a detailed recitation of the required shape of the cage, a shape that once again covers only the FIG. 1 embodiment.

Let us move along to independent claim 10:

10. A novelty drinking straw comprising:
a drinking tube defining a hollow, continuous flow passage and including first and second linearly extending end portions, one of said first and second end portions defining a mouthpiece;

a three-dimensional cage formed by said flow passage and disposed therealong intermediate said first and second end portions, said cage being formed of a plurality of horizontally and vertically extending windings to define a parallelepiped interior volume or containing a prize object dimensioned to fit therein; and

means formed by said passageway or retaining said object within said interior volume.

For the first time, we find a description in the second clause of the shape of the FIG. 2 embodiment. Of course, it excludes the FIG. 1 and FIG. 4 embodiments. Likewise, we will look at claim 11:

11. A combination novelty drinking straw and prize object contained therein, said combination comprising:
a drinking tube defining a hollow, continuous flow passage and including first and second linearly extending end portions, one of said first and second end portions defining a mouthpiece;

a three-dimensional cage formed by said flow passage and disposed therealong intermediate said first and second end portions, said cage defining an interior volume for containing a prize object having means defining a passage formed therethrough and dimensioned to fit in said cage; and

means formed by said passageway tier retaining said object within said interior volume comprising a linear extension of said one end portion which extends along said axis through said helical cage for extension through said passage of said

> prize object when said prize object is placed within said interior volume.

We find in the last clause a description of a retaining means that covers the embodiment of FIG. 4, but not FIG. 1 and FIG. 2.

Only after a thorough reading of all the independent claims of this patent can we conclude that claim 1 actually is the broadest claim. While it does not cover the FIG. 4 embodiment, it is certainly broad enough to cover straw configurations with prize cages shaped in a wide variety of ways, most of which are not depicted in the patent.

Now that you have a good understanding of what the broadest claim covers, as well as the rest of the independent claims, you would be prepared to make an intelligent determination of how the Lipson patent impacts whatever it is you propose to do, be it to manufacture a novelty straw in competition with the patented one, file a patent application on your own invention, decide whether the Lipson patent is a strong enough patent that you might want to take a license under, etc. Of course, when you are in a real-life situation, you most likely will not have to make that decision yourself. You may well take your preliminary conclusions to the intellectual property specialist in your company or to a member of a patent law firm for further advice. But your efforts in deciphering the patent will be more than amply rewarded when you find that you are able to discuss it intelligently with these specialists. They will be able to do a much better job for you because you have already done your homework.

9.5.8 Corrections

Before we move on to the next sections, which get into the meat of reading and interpreting the most critical part of a patent — the claims — we should mention one final element that is very often found in a patent: the Certificate of Correction. While the Lipson patent is blessedly free of this burden, many issued U.S. patents are printed with errors considered significant enough by the patent holder to justify stapling one or more sheets of corrections to the face of the patent. Most of these corrections involve nothing more than typographical errors, printing errors, nonstandard spelling, language found in the application that somehow got lost in the printed version, etc. Generally speaking, the corrections are much more likely to be of significance if they affect the claim language of the patent, rather than the specification. You should certainly be aware of corrections to the claim language when you read the patent. You can insert into the patent, if you wish, the corrected language, whose location is expressed in the Certificate of Correction by the same "col. a, lines x–z" convention mentioned earlier.

9.6 Understanding patent claims

Of all the parts that go into making up a U.S. patent, it is the claims that are by far the most important because they define the metes and bounds of the intellectual property that the patent protects. A useful analogy is to think of them as similar to the survey language you find in the deed to a piece of land describing exactly where on the surface of the Earth the boundaries of a particular piece of property may be found. Using that real property descriptive language, the owner can build a fence around the property to keep out trespassers. In the case of patent claims, they tell the world exactly what invention the patent protects, and the patent document, itself, serves as a kind of "fence" that keeps others from practicing the claimed invention without the patent owner's permission.

While there can be no question that claims are the most important part of any patent, they are also the most difficult for the nonexpert to understand. Problems lie in the numerous patent language conventions that have arisen over time. For example, one striking feature of a patent claim, however long and complex, is that it is always expressed in a single sentence. This mandatory one-sentence format can make reading and understanding a long, difficult patent claim a daunting task even to someone familiar with claim language. That is one reason why it is so important first to look at other sections of the patent (the Abstract, the drawings, the Summary, etc.) before tackling the claims.

You are not likely to see words like "or" or "either" appearing in a U.S. claim, particularly in older patents. Instead, you will often find very peculiar phraseology such as "one of said first and second ends." This is nothing but a roundabout way of saying "either the first end or the second end," and surely the claim language would be much clearer expressed that way. Another example of language that avoids using the dreaded "or" word is "a material selected from the group consisting essentially of: aluminum, stainless steel, case-hardened steel, thermosetting polymers, graphite fiber composites, and combinations thereof." All this has come about because the Patent Office has had a long-standing policy of strongly discouraging or even banning the use of "or" in claim language; the PTO feels that allowing its use may open the door to the abhorrent practice of "claiming in the alternative" (claiming more than one invention), strictly prohibited under U.S. patent law, or, less seriously, may render patent claim language unclear and imprecise.

As with the entire patent document, each claim has a format, and knowing that format goes a long way in making the task of reading and understanding a claim that much easier. While there are different formats of claims (as we will discuss later on), all independent claims in U.S. patent practice do share a common format. Because claims may be structured either in outline or in paragraph form (or in a combination of both), this underlying basic format is not always immediately obvious. Nevertheless, you can always find it when you know what to look for.

Each independent claim is composed of three parts: the **preamble**, the **transitory phrase**, and the **body**. Let us briefly revisit claim 6 of the Lipson patent and identify these three components within that claim. We will use bold for the preamble, italic for the transitory phase, and standard text for the body:

6. **A combination novelty drinking straw and prize object contained therein,** *said combination comprising*:

   a resilient drinking tube defining a hollow, continuous flow passage and including first and second linearly extending end portions, one of said first and second end portions defining a mouthpiece;

   a helical cage having first and second ends and a middle portion medial thereof, said cage being formed by said flow passage and disposed therealong intermediate said first and second end portions, said cage defining an interior volume and being formed by a plurality of coils wound spirally around said volume, said plurality of coils each having a circumference and being graduated in circumference from a largest circumference coil proximate said middle portion of said cage to a smallest circumference coil disposed proximate each of said first and second end of said cage; and

   a novelty prize object disposed in said interior volume, said prize being dimensioned so as to be normally retained within said interior volume and to be removable from said interior volume when adjacent pairs of coils of said cage are linearly displaced away from each other so that said object prize is passable therebetween.

9.6.1   The preamble

The preamble is usually just a short introduction that nevertheless serves some important functions. First of all, it tells us roughly what kind of subject matter the claim protects — in the case of the Lipson claim 6, the preamble tell us that what is claimed in this claim is the *combination* of both a "novelty drinking straw" and "a prize object." In order to infringe this claim, someone would have to make, use, or sell a product that includes *both* the straw and the prize object. Contrast the claim 6 preamble with that of claim 1 which merely recites "a novelty drinking straw." No prize object is recited and simply making the straw could constitute an infringement. Interestingly, both claim 1 and claim 6 go on in their respective bodies to mention "a prize object," but this does not alter the fact that claim 6 covers the combination of the straw and prize object, while claim 1 covers a straw without anything in it, provided that it is constructed so that it could include the prize object in accordance with the rest of the language of claim 1.

Here are examples of some other types of preamble language you may encounter:

- A method of measuring hydrogen sulfide concentration within a periodontal pocket
- A multi-layer laminate material suitable for use as a floor covering
- In combination, an electronic target and aiming device for shooting at the target
- A kit for retrofitting a vehicle with a child safety seat anchoring device
- A modularized assembly for transferring a workpiece sequentially through a plurality of evenly spaced, horizontally arrayed work stations
- A computerized system for selecting and booking travel arrangements for members of a sponsored group having a series of travel policy protocols covering the members of the group
- A method of manufacturing a vehicle airbag enclosure panel
- An improved ink formulation for use in an ink-jet printer
- A composition for treating sun-damaged skin
- An improved method for thermally ablating a human uterus

The preamble also serves as an introduction to the rest of the claim. It gives us a technical frame of reference into which we can fit the actual elements of the invention we subsequently find in the body of the claim. It is important to keep in mind that the structures, systems, devices, etc. that so often appear in a claim preamble may not necessarily be part of the actual invention. For example, the child safety seat anchoring kit mentioned above obviously does not include the vehicle into which it is intended to be installed.

9.6.2 The transitory phrase

Turning once again to the Lipson claim 6, immediately after the claim's preamble, we find a short transitory phrase that links the preamble to the body of the claim — "said combination comprising...." Do not let yourself be bothered by the all-too-frequent appearance of the word "said" in claim language. You can simply translate "said" as "the" because that is exactly what it means. The U.S. Patent Office is very strict about requiring that the first appearance of an element in a claim be preceded by "a" or "an." Thus, the Lipson claim starts with the words "A combination novelty drinking straw and prize object" because that is the first time the combination of straw and prize object appears in the claim. By the time we get to the transitory phrase, we have already been introduced to the combination, so we find it correctly identified as "said combination." It would have been equally correct to use the wording "the combination."

The word "comprising" simply means that the combination of straw and prize object includes at least the elements recited in the body of the claim. It may include other structures and features as well, but to be covered by the claim, it must include each and every structure recited in the body of the claim or its equivalent.

Most U.S. patent claims use the word "comprising" before proceeding to the actual list of elements. However, you will occasionally find alternative language, particularly in chemical formulation claims, "consisting of," or "consisting essentially of." "Comprising" is considered to be open-ended, i.e., while it requires all of the listed elements, it does not foreclose the addition of other, unclaimed elements. Thus, a claim for a table that recites only three legs may be infringed by one for a four-legged table (because it has three legs), provided, of course, that the four-legged version also includes all of the other elements actually recited in the claim. But a three-legged table will not infringe a claim for a table which requires that there be four legs.

If a claim needs to specifically *exclude* additional elements, then the transitory phrase will be either "consisting of" or "consisting essentially of." These appear almost entirely in chemical formulation claims where there is a need to limit the claim to the various ingredients recited. The strongest exclusionary language is "consisting of." It is completely close-ended. When it appears in a claim, it means that a composition made according to the claim must include every one of the listed ingredients and nothing more or nothing less. Less stringent is "consisting essentially of." This partially close-ended language permits additional ingredients into a chemical formulation provided that they are only incidental to the efficacy of the formulation — added colors, fragrances, emollients, etc.

Another type of transitory phrase that you will commonly encounter is found exclusively in method claims. Method claims usually employ the phrase "comprising the steps of." The method claim will then go on to list in the body of the claim the various steps required to perform the claimed method.

Since we have already introduced the concept of dependent claims earlier in this chapter, it is worth mentioning that the transitory phrases of dependent claims are slightly different in that they usually add words like "further" or "additional." Thus, "further comprising," "further including," "comprising the additional step of," etc.

In addition to smoothly transiting us from the preamble to the body of the claim, the transitory phrase has another important function; it helps us identify exactly what the claim covers. In the Lipson claim 6, for example, the transitory phrase helps us understand that it is the straw–prize object *combination* that will be defined by the body of the claim to follow. While this may be fairly obvious in a relatively simple claim like our example, it is worth bearing in mind that the preambles of some claims are long and complex, and recite a bewildering variety of structures and components.

This is particularly so if the claimed device is a subcomponent of a larger system, and the system in which it is intended to be used is described in some detail. It is easy to lose your bearings in a claim like this, and the transitory phrase can get you firmly back on the path of exactly what the claim is about.

9.6.3 The body

You will find claims that, like those of the Lipson patent, are written in an outline format, i.e., the various elements of the invention are set forth in separate paragraphs. Colons and semicolons are used for additional clarity. This arrangement obviously makes a claim much easier to read, but not all patent application drafters use this format. You may encounter claims, even quite lengthy ones, that are written as a single paragraph. When you encounter such claims, your first task will be to separate out the various elements, perhaps using underlining, numbering, or punctuation to help you visually untangle the claim. We have already learned how to recognize the preamble and transitory phrase, so you should first find and identify them. Recognizing the elements that appear in the body of the claim is not as hard as it might seem if you remember that when an element makes its first appearance in a claim, it will be preceded by either "a" or "an."

Of course, a patent claim is more than just a list of parts. In addition to the elements of the structure of the patented article, you will also encounter functional language that explains how the various parts cooperate to result in a useful device. For example, in the Lipson claim 1, we find the phrases "one of said first and second end portions defining a mouthpiece" and "said cage defining an interior volume configured to retain a prize object entirely therein." In claims covering mechanically more complex devices, you will usually encounter much more elaborate functional language. For example, a claim to a device that includes a ball valve may say something like, "said ball valve element being rotatable from a first position wherein fluid may flow through said flow passage, to a second position wherein fluid is blocked from flowing through said flow passage."

On your first pass through a claim of any complexity, it is a good idea to first make a list of the various structural components: the ball valve, the valve cage, the ball valve element, the fluid passage, the fluid inlet, the fluid outlet, etc. On your second pass, you can then look for and focus in on functional language that explains how the components work together. Having the drawings in front of you while you go through the claim elements will be extremely helpful since graphic images often convey information much better than words. But do not fall into the mistake of assuming that the claim language covers only the structures that appear in the drawings. Almost always, there will be other ways of constructing the device than what is illustrated in the patent drawings.

Watch out for what patent drafters call "means plus function" language. For example, in the last clause of the Lipson claim 1, we find "means formed by said passageway for retaining said object within said interior volume." This may seem like an odd way of defining a structural element, but its great value to a patent owner is that means-plus-function phrases do not pin the patent down to just one particular category of structure — you can say "means for fastening the bracket to the upright" and it will cover doing this with existing types of nails, screws, bolts, glue, hook-and-loop fastener, etc. For example, the Lipson patent illustrates two quite different and distinct structures that are both capable of retaining the prize object within the cage's interior volume. One type involves winding part of the straw in such a way as to create an enclosure (be it egg-shaped or parallelepiped, or whatever) to trap the prize object. The other type involves actually passing part of the straw through a portion of the prize object. Possibly, there might be yet other means for performing this particular function. Once again, do not assume that a means-plus-function phrase is limited only to the structures actually shown and discussed.

There is one additional type of phraseology you are likely to encounter in the body of a patent claim. Phrases of this type start with words such as "whereby," wherein," or "thereby" and are used to help clarify what might otherwise be confusing claim language. We find such a phrase in the Lipson claim 9: "wherein the prize object has a circumference in at least one direction sized smaller than said largest coil circumference and larger than said smallest coil circumference." A phrase starting with "wherein" is likely to provide additional specifics about the element being discussed, in this case, the relative dimensions of the prize object. Phrases starting with "whereby" or "thereby" almost always help explain the function of structure that has already been defined. For example, the following phrase could have appeared after the language quoted from the claim 9: "thereby trapping the prize object inside the cage." You will notice that this additional language would not really affect the breadth of the claim one way or another, but it does help clarify what is going on.

Of course, we know that not all patent claims cover articles or mechanical devices. There are also claims to chemical compositions, methods of making devices and chemical compositions, methods of using devices and chemical compositions, methods of medical treatment and surgery, computerized systems for performing various tasks, etc. While each of these claim categories presents its own set of unique challenges, the same systematic approach discussed above is the best way to tackle each of them.

9.6.4   Dependent claims

Reading dependent claims is usually much easier than reading independent claims because they are normally recited on one or two additional features of the invention. Always bear in mind that a dependent claim includes all the elements of any claims that it depends on, including other any dependent claims in its chain of claims, as well as the one independent claim that always stands at the head of a chain of claims.

One additional wrinkle involving dependent claims is that it is possible for a claim to directly depend on more than one other claim. For example, claim 4 of a patent may state, "The widget as described in claims 2 or 3 above and further comprising a second pair of fins disposed on a second end opposite the first end." This is an example of *a multiple dependent claim.* This effectively means that there are two variations of claim 4 — the version that contains whatever limitations are recited in claim 2 and another version containing all of the recitations of claim 3.

Multiple dependent claims are strongly discouraged in U.S. patent practice. The PTO charges a punitive fee for each one that appears in a patent application. Most U.S. practitioners avoid their usage. However, you occasionally will run across them in U.S. patents that are based on patent applications first filed abroad. They are quite common in most other countries. They do have the advantage of cutting down the total number of claims that appear in the patent, but they also make figuring out the various chains of patent claims much more difficult. If you do have to deal with a patent containing one or more multiple dependent claims, it may help to draw a simple "claim tree," with arrows linking the various chains of claims.

One thing you will not find in a U.S. patent is a multiple dependent claim that depends on another multiple dependent claim. The PTO forbids this practice. However, some foreign jurisdictions do permit it.

9.7   Claim formats

As you know by now, it is possible to patent both things (articles of manufacture, chemical compositions, computer systems, etc.), as well as processes (methods of making articles, methods of using devices, methods of treating biological organisms, computerized methods of doing business, etc.). Many patents contain a mixture of both article and method claims, such as a series of claims to a new or improved article (e,g., a floor laminate), combined with claims directed to a process for making the laminate material and/or claims directed to using the article (i.e., a special way of installing the flooring). There are also more unusual and specialized types of claim formats that you may occasionally encounter, such as a product by process claim. Let us take a closer look at some examples of claim formats you may encounter so that you will be able to recognize them and understand how and why they are put together that way.

9.7.1   Article claims

The majority of claims found in U.S. patents (and this is also true of foreign patents) are article claims, such as all of the twelve claims of the Lipson patent we analyzed in great detail in a prior section. This is not too surprising since, after all, that is what immediately comes to mind when we think of the word "invention" — a better mousetrap, a more efficient internal combustion engine, a "miracle" spot remover, a drug to combat AIDS, and on and on, as far as our imaginations will carry us.

For example, let us suppose that no one had ever thought of the idea of a pencil up to now, and finally some inventor has a brainstorm and decides to patent his wondrous new device. Here is an example of a set of possible claims for a patent application:

1. A writing implement comprising:
    a tubular length of marking material; and a generally cylindrical jacket disposed around said marking material so as to leave at least a first end thereof exposed for marking a writing surface.
2. The writing implement of claim 1 further comprising a body of erasure material disposed on a first end of said implement for erasing marks made by said marking material.
3. The writing implement of claim 1 wherein a second end of said implement opposite said first end terminates in a frustoconical point formed by removing portions of said marking material and said jacket.
4. The writing implement of claim 1 wherein the marking material is formed of carbon.
5. The writing implement of claim 1 wherein the jacket is formed of wood.
6. The writing implement of claim 2 wherein body of erasure material is formed of rubber.

While these claims are simpler than the claims of the Lipson patent, you will notice that they have much in common. Claim 1 contains the preamble, transitory phase, and body discussed earlier. The body of claim 1 is a recitation of the two main structural elements — the length of marking material and the enclosing jacket — that make up the pencil and also describes how they fit together to form a single article. We also see mention of a third element — the marking end — that is not exactly a separate structural element, but is formed by the way the two main elements are arranged with respect to each other. We also see a functional limitation in this claim — that the exposed end is there for the purpose of enabling us to write with the device.

The first of the dependent claims, claim 2, adds the familiar eraser to the pencil. Note that the device of claim 2 also includes everything recited in claim 1 — the length of marking material, the jacket, and the exposed writing end. Claim 3 adds something else very important to the user of the pencil. Claim 1 is broad enough to cover a pencil with a blunt, unsharpened end — not a very useful article, but typically the way that pencils are sold in commerce. Claim 3 adds the further limitation of tapering the end opposite the erasure by removing portions of the marking material and jacket to give us what is, after all, the point of a pencil. Now we have a claim that covers all sharpened pencils.

The other three dependent claims merely recite some preferred materials for forming the various structural elements of the pencil. This type of dependent claim is commonly found in a chain of article claims. Particularly note that claim 6, which defines the erasure material, depends on claim 2 and not on claim 1. It cannot depend on claim 1 because no eraser is recited in claim 1.

Sometimes, in reading article claims, you will come across a term with which you are not familiar, or perhaps a familiar term used in an unfamiliar way. It is not unusual for inventors to actually coin a term for a particular element and use it in the patent application. However, the specification should always contain definitions of terms like this so that anyone reading the patent can understand exactly what they mean when they are used in a claim. One fast and convenient way to find these explanatory passages in the specification is to do a keyword search for the mysterious term in the on-line text version of the patent you can find at the PTO website (www.uspto.gov).

The class of article claims is broad enough also to encompass claims to compositions of matter. Typically, a composition claim will include a list of constituent ingredients ("A composition for treating acne comprising: alphahydroxy acid; limonene; and a pharmaceutically acceptable carrier"), often with ranges ("said limonene being in a range of 5 to 10 weight percent"). Frequently, dependent claims further narrow the ranges. The constituent ingredients may be identified by their common names, by chemical nomenclature, by how they are prepared, by their total molecular weights, by their mass spectrometer fingerprints, etc. The specification in a chemical composition case should describe the nature of each constituent ingredient in enough detail to enable one of skill in that particular chemical art to either procure it or prepare it. Frequently, chemical cases will identify constituent ingredients that are in commercial use by their brand names.

9.7.2   Process claims

This is probably the second most common category of claim. Process claims include methods of making or manufacturing articles, types of medical treatment, ways of processing data, purifying and refining raw materials, handling and transporting articles of manufacture, methods of doing business, methods of transmitting electromagnetic waves, etc.

Process claims have their own format. To illustrate this format, let us look at a simple hypothetical example — a method of making a familiar culinary delight:

1. A method of making a grilled cheese sandwich comprising the steps of:

    coating a first side of each of two slices of bread with softened butter so as to form a thin butter layer thereon;

    cutting a body of cheese into at least one slice dimensioned to be received between said slices of bread;

    coating a flat metal surface with a nonstick agent;

    heating said surface to approximately 325 degrees F;

    placing said at least one cheese slice between said bread slices so as to form a sandwich, said butter layers being left exposed;

    placing said sandwich on said heated surface such that a first exposed, buttered surface thereof is in contact with said heated surface;

    cooking said sandwich until said first exposed surface begins to brown;

    turning over said sandwich to bring a second exposed, buttered surface in contact with said heated surface;

    continuing to cook said sandwich until said second exposed surface begins to brown and said cheese begins to melt to form a cooked sandwich; and

    removing said cooked sandwich from said heated surface.

2. The method of claim 1 comprising the further step of cutting said cooked sandwich transversely in half.

You will notice that the details of the claimed method are listed as a series of steps. Does the claim require the steps to be performed in any particular order? Yes and no. Certainly, some of the steps must be performed after others. It would be impossible, for example, to place the sandwich on the heated surface before you've gone through the steps required to make the sandwich in the first place. On the other hand, there is no reason why you could not heat the pan before proceeding to assemble the sandwich. Just because the steps appear in the order they do does not necessarily mean that they must be performed in that particular order.

Another thing to note is that the components that make up the final product change as the grilled sandwich is gradually created, and the language used to describe the changing components also changes as the process moves to completion. For example, we start out with "two slices of bread," which then each acquire "a thin butter layer." When the sandwich is assembled, the surfaces of the bread slices carrying their respective layers of butter become "exposed surfaces," and the butter, bread, and cheese components turn into a "sandwich," which eventually becomes "a cooked sandwich." This changing language helps us figure out the order in which at least some of the steps must be performed.

Occasionally, you will find steps in a method claim that start with words like "first" or "subsequently." Time-dependent terms like these actually do fix the order of the steps of the method, so be careful to watch out for them.

Claim 2 is an example of how a typical dependent method claim is formatted. The additional step or steps are described in the same orderly fashion as the steps of the base claim. In order to practice the method of claim 2, you would have also to perform all the steps set forth in claim 1.

Sometimes you will run across dependent claims in a string of method claims that do not recite any additional steps, but further define an article or substance used in practicing the base method. For example, we could add a claim 3 to the above string — "The method of claim 1 wherein the said flat metal surface is a frying pan." Strictly speaking, this kind of dependent claim is not considered best practice because it really amounts to modifying a method claim with a material limitation. Some patent examiners will object to such language in a method claim. A more orthodox way to say the same thing would be, "The method of claim 1 further comprising the step of [or "comprising the further step of"] providing the flat metal surface as a frying pan."

9.7.3 Method of use claims

Another broad category of claims commonly found in patents are called "method of use claims." These always involve starting with an article, device, substance, software program, etc. and then using or applying it in some fashion to achieve some desired result. Often you will run across method-of-use claims in the same patent in which you find article claims to the same invention that is used in the method claims.

Let us revisit the invention of the pencil we discussed earlier. Claim 1 to claim 6 are all, of course, article claims. But surely there is more to the inventive concept of the pencil than just its mere structure. After all, when the light bulb goes on for any inventor, questions that will be foremost in his or her mind are "What is this invention good for? What problem does it solve? What are people actually going to do with it, and exactly how are they supposed to use it?" The concepts of an article of invention and how it is used are so intimately bound up that no patent specification would be complete unless it answered these questions.

In the case of the pencil, doubtless any patent directed to it would fully describe its method of use — cutting away part of one end to make a sharp point, grasping the device, bringing the sharpened end into contact with a writing surface and applying enough pressure to leave a mark or series of marks on it, etc. An exemplary claim string you might find in our hypothetical pencil patent might be:

7. A method of marking a writing surface comprising the steps of:

    forming a marking implement by:

    >   providing a tubular length of marking material: and

    >   disposing a generally cylindrical jacket disposed around said marking material so as to leave at least a first end thereof exposed for marking a writing surface;

    sharpening a first end of said marking implement by removing portions of said jacket and said marking material so as to form a frustoconical point;

    grasping said marking implement proximate said first end;

    bringing said point to bear against said writing surface: and moving said point along said writing surface such that a visible trail of marking material is deposited on said surface.

8. The method of claim 7 comprising the further steps of:

    forming an eraser from a body of erasure material;

    disposing said eraser on a second end of said implement opposite said point; and

    rubbing said eraser over said visible trial of marking material until said trail is no longer visible.

9. The method of claim 8 comprising the further step of forming said body of erasure material of soft rubber.

Now we have added to our original article claims covering the pencil a string of claims directed to using a pencil to write with, as well as to erase with.

9.7.4   Product by process claims

Before we leave the subject of claims formats, there is another type of claims commonly found in U.S. patents with which you should have some familiarity. A product by process claim is simply a claim that covers an article or substance made by a particular inventive process or method. You most commonly encounter such claims as additions to what are basically process or method patents.

For example, let us take a further look at the grilled cheese sandwich. The string of claims we previously discussed covered a method of making such a sandwich. Doubtless, one could think of other methods — putting it under a broiler, grilling both sides at once in a closed grill, toasting it over a campfire, etc. We might skip the step of buttering the bread entirely. We could use cheese spread instead of sliced cheese. And so on. None of these variant ways of making a grilled cheese sandwich are literally covered by the claims above, and it is unlikely that performing any of them would infringe the patent. So no one could argue that simply inventing one method of making such a sandwich gives anyone the right to claim they have invented a grilled cheese sandwich.

While that is true, there is a practical problem with simply limiting what is essentially a method patent to strictly method claims. Method claims are more difficult to enforce than article claims. Infringing a method claim requires that someone actually practice the method. This means a patent owner has only the manufacturers or makers to target in an infringement action. But article claims can be infringed by making, using, selling, or offering to sell the patented article. Now the patentee has a whole gallery of parties to target in the infringement suit — the manufacturers, the distributors, the wholesalers, the retailers, the users, etc. So having a product by process claim in a patent can give a patentee a considerable advantage when it comes to enforcing the patent. Such a claim typically appears in the following format:

3. A grilled cheese sandwich produced according to the method of claim 1.

However, some examiners will not permit claims of this sort, so you also might see something like:

4. A grilled cheese sandwich produced by the method of:

coating a first side of each of two slices of bread with softened butter so as to form a thin butter layer thereon;

cutting a body of cheese into at least one slice dimensioned to be received between said slices of bread;

coating a flat metal surface with a nonstick agent;

heating said surface to approximately 325 degrees F;

placing said at least one cheese slice between said bread slices such that said butter layers thereof are exposed so as to form a sandwich;

placing said sandwich on said heated surface such that a first exposed, buttered surface thereof is in contact with said heated surface;

cooking said sandwich until said first exposed surface begins to brown;

turning over said sandwich to bring a second exposed, buttered surface in contact with said heated surface;

continuing to cook said sandwich until said second exposed surface begins to brown and said cheese begins to melt to form a cooked sandwich; and

removing said cooked sandwich from said heated surface.

Like method-of-use claims, product-by-process claims are of mixed format. In fact, they are sort of a mirror image of method-of-use claims. A method-of-use claim is basically an article claim slightly altered to put it in a method format. A product-by-process claim, in contrast, is basically a method claim slightly altered to put it in article format. Recognizing these mixed format claims when you see them and having some understanding of why they are used can only be of benefit to you in your patent reading efforts.

9.8   A final caveat

Armed with the information this chapter provides, you should now be able to find your way through any patent without undue difficulty. In fact, you may be able to learn whatever it is you need to know quite quickly, simply by using the systematic approach outlined above. On the other hand, the answers you want may not be so obvious. For example, even after carefully reading the entire patent document, making claim charts, and looking up difficult terms, you still may not be able to decide whether the new product you want to make infringes the claims of the patent.

Some patent issues are simply too complex for one person to try to figure out, even from a careful study of a patent. After all, questions of patent infringement can result in many years of litigation, employ scores of attorneys, and consume millions of dollars in fees and expenses. Patents are legal documents and subject to conflicting legal interpretations. In that way, they are no different from a deed to a piece of real property, a contract to convey title, a franchise agreement, etc., and we all know how lengthy and complex disputes involving any of these types of documents can be.

At some point, you may need to seek expert advice. Particularly if there is a question of patent infringement hanging in the balance, you definitely should take your preliminary conclusions to a patent attorney for an expert opinion. Even if that necessity comes to pass, your efforts in patent reading and claim interpretation will be anything but wasted. After all, the more you know about the patent in question, the more you will be able to learn from others.

chapter ten

Technology transfer: patent licensing and related strategies

Peter J. Newman
University of Alabama at Birmingham

Contents

10.1   Introduction

After a patent has been applied for or awarded, a challenge presents itself that is similar to the question of why dogs chase cars. Namely, a plan should already exist regarding what will be done with the object being sought when it has been acquired [1]. This chapter will provide one such plan for commercializing patents with an overview of practices that are commonly used in the field of technology transfer. Some of these practices relate directly to licensing patents. The chapter also outlines the broader technology transfer infrastructure, such as forming start-up companies and negotiating interinstitutional and materials transfer agreements.

There are several ways to benefit from patents. One option for a company is to utilize the invention to develop and sell products based on one or more claims of the patent. In addition, enforcing the patent rights lessens the threat of competition in the marketplace by blocking others from making, using, or selling products based on the patent claims. Alternatively, patent rights may be conferred upon one or more recipients in return for compensation. This granting process is part of the field known as **technology transfer**, and the space through which partially developed technology is moved toward product development and eventual commercialization has been referred to as the **technology transfer gap** [2].

Broadly, technology transfer can be defined as "the range of processes, from knowledge transfer to product or process commercialization, that facilitate the development and adoption of knowledge and technology" [3]. This chapter, however, focuses on bridging the technology transfer gap by granting commercialization rights to patents and associated know-how to an outside party via a **License Agreement**. Technology transfer in this narrower sense is commonly known as **licensing** and consists of the movement of technology from the seller (the **licensor**) to the buyer (the **licensee**) via a legal document granting the licensee permission to use the technology claimed in the patent (the **License Agreement**) [4]. Usually such a transfer involves granting patent rights in return for royalties and other financial consideration, but there are cases where the remuneration may take the form of a trade of rights, such as a **cross-license**, or even be done solely as an obligation to the government for its sponsorship of the invention.

Transferring technology through patent licensing dates back at least to the 19th century. As an example, George Sheldon's U.S. "road engine" patent, with a priority date of 1877, was licensed to early American automobile manufacturers in exchange for royalties. In a noteworthy confrontation, however, Henry Ford chose to fight (successfully) the patent's validity in court rather than negotiate a license agreement [5].

10.2 Purpose of technology transfer

Technology is transferred for several reasons. Sometimes, e.g., at a university, the primary goal may be to make the technology and resulting products available to the general public by out-licensing the patent. At other times, e.g., at a company, it may be done to avoid costly patent litigation whereby cross-licensing is the logical business solution. Patents may also be donated to a nonprofit organization for charitable or tax reasons. However, most technology transfer involving licensing patents is done with the intent of making money. The licensor typically is hoping to receive **royalties**, which are a share in the proceeds generated from sale of products or services based on the invention. The licensee usually is hoping to develop and sell products or services based on the in-licensed patent rights. As a consequence of purchasing a license to legally infringe the licensor's patent rights, the licensee has access to technology unavailable to nonlicensees. Nonlicensees that utilize the same technology claimed in the patent are known as **infringers** and, if caught, must cease the patent infringement, acquire a license, or else successfully challenge the patent's validity in court.

In addition to generating licensing income for the inventor's employer, technology transfer may benefit the innovators themselves in different ways. At a company, the inventors often receive recognition from the firm that may include a modest financial reward, although this typically excludes any money received directly from out-licensing the patent. At a university or other nonprofit research institution, inventors are usually rewarded according to a **patent policy** that states how licensing income will be divided and distributed. The patent policy generally states that a portion of any licensing income must be distributed to the inventors and that another portion goes to those departments and centers that provided the space, equipment, and financial support for the invention. An independent inventor, who has assumed the risk and provided the financing for making the invention, keeps all the income received from licensing the patent in the absence of a different arrangement, such as the use of a technology broker.

Licensing, in contrast to assigning, a patent also permits retention of title, i.e., ownership, to the patent as well as the potential for partitioning the grant of rights by degree of exclusivity, geographic region, or field of use. Retaining title is also important because it is easier to get the patent rights returned in the event a licensee violates the license agreement or goes bankrupt. Furthermore, a license agreement usually entitles the licensor to a stream of royalty income over many years rather than a single lump sum, which may be preferred for tax reasons.

Often, particularly at nonprofit institutions such as universities and government laboratories, the technology being transferred arises from basic research and is therefore at a very early stage with regard to product development. For example, when a university is ready to license its inventions to industry, most have not even reached the prototype state, much less shown manufacturability and practicality in the marketplace [6]. Consequently, the licensee of such embryonic technology is bearing a greater risk than would be the case with technology that is already proven in the marketplace. From the licensee's perspective, technology transfer provides tools to pursue the rewards of bearing this risk.

In the case of inventions arising from grant funding provided by the U.S. Federal government, technology transfer is also done as a legal obligation. While certain segments of the U.S. Federal government, such as the National Institutes of Health, permit funded institutions to own and license resulting inventions, the government is also entitled access to such inventions. The Bayh-Dole Act of 1980 (PL 96-517) permits institutions to own innovations made with Federal money, with the understanding that an effort will be made to patent and license the inventions so that new products and services become available to the public. A major goal of the Bayh-Dole Act is to provide an incentive for the commercial sector to invest in technologies arising from government-funded research. Consequently, this is expected to create jobs and stimulate the economy, both locally and nationally.

For the licensor, technology transfer can generate income, particularly when licenses are granted to appropriate partners and synergistic patents are grouped together under the control of a common licensee. To make a simple analogy using the game Monopoly®, if patents are viewed as the different-colored properties, technology transfer is similar to the process of making deals to group properties of the same color together to optimize financial gain. The colored properties by themselves are of relatively little value, just as a single patent typically derives most of its value through synergistic combination with other intellectual property.

From the licensee's perspective, using technology transfer to facilitate the acquisition of beneficial technology without having to invent it from scratch can save considerable time and money. For example, after World War II, Japan became a leading producer of advanced technical products by adopting technologies that had been developed in western countries. Japanese companies entered into licensing agreements with many of the leading firms in particular industries [7].

10.3 Finding potential licensees

A number of resources exist that are useful when seeking potential licensees. The U.S. Patent and Trademark Office Web site allows a search to find the assignees of issued U.S. patents in the same class and subclass as the invention.[1] Similarly, some U.S. patent applications can be viewed if they

---

[1] The USPTO Web site is currently found at *http://ftp.uspto.gov/patft/index.html*.

were filed after the date the U.S. changed its policies concerning the publication of pending patent applications (March 15, 2001). Search for pending patents that have been published in Europe at the European Patent Office Web site.[2] The owners or licensees of issued and pending patents relevant to the same markets as the invention may be the companies most likely to benefit from the patent rights. Additional resources for finding licensees are directories that categorize companies by their product development interests or product classifications.[3] An Internet search can be an effective tool since many technology companies have home Web pages describing some of their ongoing research and development efforts.

To learn about companies that may be suitable licensees, become educated with regard to the invention's relevant markets. Identify companies that either offer products in these markets or else are in the process of developing such products. Search worldwide, with an emphasis on firms positioned to compete successfully in countries where the invention is most applicable. Next, obtain information concerning the approximate market share for each of the companies.[4] What firms are competitors in these markets? What are each company's strengths and weaknesses? Investigate the laws pertaining to regulation of companies in a particular field, such as Food and Drug Administration requirements for pharmaceutical development and sale.

Think broadly when considering what markets apply to the invention, perhaps even beyond the obvious markets that were conceived of when the patent application was first filed. Sometimes, a firm may infringe patent claims in a field completely different from what was first envisioned by the licensor. For example, a radiology patent with medical utility initially had its broadest market in airport baggage scanning equipment.

Estimate the size of the markets. This information is often available at the Web sites of nonprofit societies dedicated to providing information about a particular problem, such as a disease, that can be affected in a positive way by the invention. Other useful market-share resources can be found at the Web sites of some university business schools.[5] Alternatively, the annual reports of public companies may show product sales volume which, when combined with similar information from major competitors, can yield an estimate of market size. This is useful for assessing which types of companies are likely to put significant resources into developing products utilizing the invention. For example, if the product is a pharmaceutical or biotechnology product, is the market small enough that the product might qualify for

[2] The European Patent Office Web site is currently found at *http://ep.espacenet.com/*.
[3] Two such examples are the CorpTech Directory (published by OneSource Information Services, Inc.) and the Thomas Registry (published by Thomas Publishing Company).
[4] This can be found through the Internet, e.g., at *http://www.marketresearch.com* or *http://www.lib. duke.edu/reference/subjects/business/m_share.htm*, both of which offer helpful resources. The Web pages of firms in a particular market may also cite this information.
[5] For example, *http://www.lib.duke.edu/reference/subjects/business/m_share.htm*.

orphan drug status? If so, a smaller biotechnology company might be more interested than a large pharmaceutical firm.

An important consideration when identifying potential licensees is to find companies that will devote significant resources toward developing the invention into a product with a sustainable competitive advantage in the marketplace. This often requires finding firms that already have a budgeted research and development program in the field of the invention. These might be companies that currently have products on the market and are trying to maintain or improve their position through "next generation" products. Or, they might not yet have any actual products in the particular field. As stated previously, a good way to find such companies is to search the Internet, where many companies will show products under development at their Web site. Additionally, look for companies advertising job openings in a particular field listed in a scientific or trade journal. Many companies also routinely send out letters indicating areas of current licensing interest to the membership of technology transfer professional organizations.

Critique the invention from a prospective licensor's point of view. For example, does the patent fill an important gap in a wall of intellectual property that the company controls? Will this patent "wall" surround and provide a sustainable competitive advantage for products with significant sales potential? If not, the technology may need further development, along with an accompanying continuation-in-part application after additional inventive contribution.

As with other assets, the value of patents and other intellectual property increases with rising demand. Consequently, a logical goal is to find well-funded licensees that have strong financial incentives to acquire the patent rights and accompanying know-how. Perhaps the patent is blocking the path to product development for several firms. If this is the case, a decision will need to be made whether to grant them all licenses in the same field and territory (**non-exclusive licenses**) or grant just one that, as a premium for exclusivity, might pay significantly more to be the sole licensee (an **exclusive license**).

Look at the territory and exclusivity considerations relevant to licensing the patent. For example, in what regions of the world is the invention most applicable? This particular marketing and licensing consideration should be made in concert with decisions concerning international patent rights such as designating PCT countries and national phase filings. It is often not necessary to grant rights for all applications of the invention in order to sign a lucrative licensing deal. For example, patent rights can be licensed by apportioning to different licensees different geographic territories or **fields** of use, thereby increasing the utilization and money-making potential of a single patent. The field of a license defines the boundaries wherein a license can be practiced, such as "cancer diagnosis." The geographic territory might state, e.g., "the United States and Canada." As a general rule, however, one should grant rights only to areas where the licensee has plans to develop and the necessary resources to effectively utilize the licensed rights to develop products. Think of the previous Monopoly® analogy, where players can only build on properties where they hold all of the same color and, even then, must have sufficient money with which to buy houses and hotels for those properties.

Often, either the patent owner finds the licensee or a third party (such as a technology broker) enables them to find each other. Sometimes, however, a prospective licensee will contact the assignee after the patent has been issued in the U.S. or another major market country. This often leads to the most lucrative licensing deals for both the licensor and the licensee. In these instances, the patent is typically blocking the company's product development or is otherwise vitally necessary in some way (see Example 1 at the end of this chapter).

Finally, from a licensing standpoint, it is important for the patent attorney and the inventor to draft the patent application with the task of marketing and licensing the invention already in mind. Ideally, claims are written that are not easily designed around and where any infringement will be detectable [1]. Most inventors, however, are not skilled in the art of writing patent claims. The unfortunate consequence of this is that, although patent attorneys typically ask their client whether they agree to narrow their claims, many inventors do not understand the consequences of such a limitation [8]. To illustrate, a patent application was narrowed from claiming the inhibition of several common medical problems to one particular embodiment of the invention for treating a single uncommon disease. This became a problem when potential licensee companies were sought, but were primarily interested only in the broader invention.

10.4   Contacting potential licensees

After potential licensees for the patent have been identified, determine how best to approach them. Begin by writing a one-page document or flyer stating the applications and advantages of the invention in addition to the current patent status. The flyer should contain only nonconfidential information and be written to solicit interest from prospective licensees. The title is particularly important since the persons who will review the flyer are often inundated with similar materials through the mail and over the Internet. Therefore, the title should be strikingly relevant, and keywords or phrases that are used by target companies when describing their markets should appear in the title.

If the patent has been published in Europe or the U.S., a corporate contact may also be referred to the appropriate WO number through the World Intellectual Property Organization. A published patent application is no longer confidential and a company can therefore obtain it freely without first signing a confidentiality agreement. This is helpful when approaching potential licensees since companies may be hesitant to be bound by such an agreement so early in the process of evaluation.

Next, select an appropriate initial contact person within the firm. For a relatively small company, this might be the president or chief executive officer (CEO). With a larger company, the Director of Business Development for the appropriate corporate division is a suitable point of contact.[6] Additionally, it is highly desirable to have a scientist who is willing to "champion" the technology within the company by advocating to the licensing office or applicable research division the invention's likely benefits to the firm. Therefore, a good initial contact at a potential licensee of any size is also a colleague scientist who can effectively influence the firm's management with regard to in-licensing patents.

Several technology transfer professional organizations publish directory listings of their members and hold conferences providing opportunities to network with both licensors and licensees. Three of these are the **Association for University Technology Managers** (AUTM), the **Technology Transfer Society** (T2S), and the **Licensing Executives Society** (LES). A large part of AUTM membership consists of technology transfer professionals who work at nonprofit research organizations such as universities and government laboratories [9]. T2S is a nonprofit organization that promotes itself as being "dedicated to sharing methods, opportunities, and schools of thought with the technology transfer community" [10]. LES, and its parent organization, Licensing Executives Society International (LESI) exist to facilitate and educate stakeholders about the profession of licensing.[7] All three of these organizations are good sources of contacts at potential licensee companies.

10.5 Technology brokers

Another resource for licensing patents is the use of **technology brokers**. Technology brokers are firms that will perform some or all of the technology transfer responsibilities, and pay the associated expenses, typically for a percentage of any licensing income.[8] Such firms will either exclusively license or take title to a patent while doing the work and absorbing the expense of ongoing patent costs and seeking licensees for the patent.

[6] Sources for locating these contacts include resource guides to companies, e.g., CorpTech Directory[R], as well as the company's Web site.
[7] LES also cites a list of rules of ethical conduct for LES members. Accordingly, the following are areas where ethical lapses may be sufficient for dismissal from the organization: (1) Compliance with Laws and Regulations, (2) Obligation under other Rules of Ethics, (3) Misrepresentation, (4) Conflicts of Interest, (5) Interest in the Subject Matter Being Negotiated, (6) Confidence, (7) Advertising and Solicitation, and (8) (unauthorized use of) Membership Lists [20].
[8] Typically 40% or more.

Brokers often ask that title to the patent or patent application be assigned to them so that they become the owners and can directly license, rather than having to sublicense, the invention. The broker then sends payments to the inventor according to a revenue sharing agreement.

Another potential reason to use technology brokers is that they can, in addition to providing ongoing patent costs and licensing experience, also provide "proof of concept" money to further advance the invention, making it more valuable and licensable. One such technology broker refers to this focus on finding embryonic or early-stage technologies and moving them through the needed development as the "venture gap" [11].

10.6   Confidentiality agreements

After potential licensees have reviewed all of the publicly available, non-confidential information about the invention such as a marketing flyer, journal reprints, and issued patents, if still interested, they will likely also wish to see proprietary and confidential documents pertaining to the invention. These include patent applications and unpublished data and know-how. At this point, a **Confidential Disclosure Agreement** (CDA), also called Confidentiality Agreement, Nondisclosure Agreement, or Secrecy Agreement, will need to be signed by the parties. A CDA is an agreement between two or more parties agreeing to keep certain disclosed information confidential, and therefore not to disclose it to other parties. A CDA provides protection against the consequences of an unprotected disclosure of confidential information, such as theft of intellectual property or loss of patentability.

Signing a CDA prior to sending proprietary and confidential information about the invention protects both parties in two important ways. First, in cases where a patent application has not yet been applied for, it protects against disclosing information that might otherwise eliminate the novelty and nonobviousness of the invention. Second, regardless of whether or not patents are already pending, a CDA protects both parties from the danger that the information disclosed might become incorporated into a patent application of the other party [12].

Typically, a CDA will have a clause specifying the circumstances whereby disclosed information either becomes nonconfidential or is considered to be nonconfidential from the start. When negotiating a CDA, it is often best for both the provider and the recipient of confidential information to insist in this section that oral disclosures of information be reduced to writing. This creates a tangible record of the exchange of confidential information that will likely make the agreement easier to enforce.

In a particularly unfortunate case, a scientist was convinced by a small firm to be flown to their corporate headquarters to discuss his most recent findings, presumably with the understanding that research and consulting agreements would later be put in place. However, since the meeting took place before a confidentiality agreement was signed, the inventor's employer had little recourse when a subsequent patent application filed by the company contained many of the same inventive findings that the scientist had divulged. This was particularly problematic since the company was doing research in the same field and, furthermore, a patent application had not yet been filed on behalf of the inventor.

10.7   Start-up companies

Another way to profit from owning a patent is to create the licensee by forming a new company that is developed around a license to the patent rights. The primary advantage of licensing to a **start-up company**, as opposed to an established firm, is that a start-up licensee is more likely to permit the licensor to become a part owner of the licensee, i.e., to receive stock or other equity in the company, as part of the compensation for licensing the patent. This is often advantageous to the licensor because, even if the licensed invention does not wind up being part of a marketed product, the owners of the firm benefit from all of the company's products and financial success. Signing an equity-bearing license with a start-up company enables sharing in the upside potential of all these future products, whether or not they are derived from the licensed patent that was paid for, in part, by the equity stake. Furthermore, to illustrate the potential for economic development, one university found that about 77% of the investment in that university's technology and 70% of the jobs created were associated with start-up companies [6].

An additional reason to consider licensing to a start-up company is that it might more aggressively develop the invention into products and be less likely to "shelve" the technology than might be the case with a large, well-established firm. Shelving the patent will likely mean that little remuneration is seen after the invention is put aside, "on the shelf," in favor of other products or alternative technology. Although shelving can be reduced by the inclusion of development milestones and annual minimum payments in the license, large companies usually have many products on the market and under development and are therefore more likely to have internal competition to the patent.

Start-ups also have several aspects that might make them undesirable as licensees. Beyond any potential liability to the shareholders in the event of litigation, lower license closure fees, or guaranteed "up-front" payments, are typically called for in start-up licenses. There is also typically a greater risk of the licensee going bankrupt during the term of the license. Additionally, a start-up might not have the financial resources to fund product development all the way to an actual product in the marketplace without the assistance of a large company partner.[9] Last, unless the licensed invention is sufficiently broad to enable a series of products to be developed from it, the start-up may be a company living or dying on the success or failure of a single-product concept. Thus, while licensing to a start-up can ultimately have a diversifying effect, initially at least, it can also result in "putting all your eggs in one basket," depending on how the company is managed.

Key issues to keep in mind when participating in the formation of a start-up company include ensuring that skilled management is hired to run the company and that the management team is equipped with a well-conceived business plan to facilitate the acquisition of capital to support the company. A **platform technology**, i.e., one that can be used to make multiple types of products, is often a good candidate to be licensed to a start-up, particularly if the technology is still embryonic and unrefined (see Example 2 at the end of this chapter). The reason for this is the logic of diversification and the ability to fund some applications internally while others can be out-licensed as a source of capital.

When taking equity in a start-up, several additional issues need to be considered. For example, will the licensor have voting representation on the board of directors? Under what circumstances can the equity be sold? What, if any, antidilution protection will the equity-holding licensor possess? Issues such as these are often addressed in a separate **Stock Purchase Agreement** or Shareholders' Agreement, and can make licensing to a start-up more complicated and challenging than licensing to a more established company.

One route that can be helpful when the start-up's scientific founder is not already an experienced business manager is to partner with a **venture management** firm. Such a firm can provide immediate management and an experienced board of directors for the company. From a financing perspective, it may also serve as a very early-stage, very "hands-on" **venture capital** firm [13]. A venture management company may partner with the inventors and be willing to provide the early-management structure for the start-up, often receiving founders' stock or other equity in the start-up in return for its services. The presence of an associated venture capital fund, i.e., a firm that invests in early-stage companies, can make the early fund-raising for seed capital an easier task.

Some fledgling companies also find it helpful to grow for their first few years in a **business incubator** facility. In such an incubator building, start-up companies are housed in a common facility and have the opportunity for frequent contact with employees of other young companies as well as attractive rent and lease rates. Further cost savings may be realized through sharing of equipment and business service providers, such as accountants, lawyers, and insurers. Business incubators are often owned by a city or

---

[9] This is particularly true with biotechnology and pharmaceutical products where R&D and clinical trials for a single drug can cost several hundred million dollars.

state, e.g., through a state university or other stakeholder with an interest in the economic success of a geographical area, and may be situated in close proximity to other, more established, companies forming an aggregate **research park**.

To supplement other sources of funding for a young technology company such as venture capital or wealthy "angel" investors, the U.S. Federal government has a grant program called Small Business Innovative Research (SBIR). A sister program that permits a majority of the funds to go to a collaborating nonprofit institution, such as a university, is called Small Technology Transfer Research (STTR).[10] Grants such as these are typically very attractive to start-ups, and to small companies in general, because they provide financing that does not dilute the ownership interests of the current stockholders.

10.8   Types of commercialization agreements

The technology transfer agreement that is used to grant patent rights to another party is a license agreement. Sometimes, however, a potential future licensee may be unwilling to commit to the expenses and obligations that accompany signing a license agreement. They may, nevertheless, want to ensure that other firms do not acquire the patent rights during their evaluation. This type of commercialization agreement is called an **Option Agreement** (or, occasionally, a "standstill agreement"). An option agreement grants another party the right to obtain a license (typically exclusive) within a stated period of time, called the "option period" or "option term" (see Example 3 at the end of this chapter).

One reason a company may wish to option a patent is that an option is typically less costly than a license. The optionee can prevent competitors from obtaining rights to the patent while deciding whether or not it will be economically desirable to proceed on their part with obtaining a license. This entails less commitment from the prospective licensee, but the benefits are also less since an actual license is not being granted.

From the patent holder's perspective, it is important to limit the option field strictly to the patents that are under consideration for licensing. To illustrate this point, an option field was written significantly broader than a single optioned patent to cover, instead, an entire area of scientific pursuit. This became a problem when an additional patent application was filed on behalf of a different inventor at the same institution and this new invention fell inside the broad field that was already optioned to the company.

Sometimes, the patent holder and the commercialization partner may wish to collaborate to further develop the invention. The commercialization

[10] The Web site for SBIR and STTR funding is currently found at *http://www.sbaonline.sba.gov/ sbir/indexsbir-sttr.html.*

agreement covering this arrangement may be called a **Cooperative Research and Development Agreement** (CRADA). A CRADA is commonly used when one of the parties is the U.S. Federal government but can be used between the private sector and universities and between two companies as well. CRADAs are signed when further development by both parties is necessary prior to commercialization. In cases where the agreement covers research being done solely at one institution, however, it is more commonly referred to as a **Sponsored Research Agreement** (SRA). SRAs are common between universities and industry, whereby the company provides funding for a project that originated at the university, e.g., in exchange for an option to license any resulting inventions.

Where patent rights are assigned to more than one entity, usually the result of each party having one or more inventor on the same patent, an **Inter-Institutional Agreement** (IIA), is often signed. An IIA states how patent costs and licensing income will be divided, as well as which party will be primarily responsible for marketing and licensing the patent on behalf of all of the assignees.[11] In the absence of an IIA, however, each assignee has an equal right to nonexclusively out-license the entire patent. Thus, it is important to name one party as having the right to negotiate licenses on behalf of all the assignees to avoid a perception of reduced value resulting from the same patent being available for license from multiple sources. Having a single "lead party" also keeps alive the possibility of exclusively licensing the patent, either to a single licensee in all fields or to multiple licensees in different fields.

The most common reason for signing an IIA is the existence of more than one assignee on a single patent. An institution without any inventors named on the issued patent will typically not be a party to an IIA and thus will not be entitled to any subsequent licensing income. There are instances, however, where contributions made solely to the specification of a patent can, nevertheless, justify the necessity of such an agreement. Identifying and profiting from such situations is part of the art of technology transfer.

An IIA may also be useful when two licensees of the same patent holder collaborate with each other and there arises a danger of a resulting invention falling within the licensed field of both licenses. In this situation, an IIA can be signed that preapportions rights to any such future patents. This limits the potential for unintentionally breaching one or more previously signed agreements as a consequence of "double licensing" the future invention.

Often, technology transfer collaborations require that physical materials be sent from one party to another, above and beyond the licensing of intangibles, such as intellectual property. This type of technology transfer arrangement, which can also be a prelude to a more substantial relationship between the parties, is generally covered under a **Materials Transfer Agreement** (MTA). Although an MTA is a form of nonexclusive license,

---

[11] In other agreements, these issues are often addressed in a section titled "Joint Inventions."

substantial fees are rarely paid by the recipients of the materials since they typically wish to use them solely for noncommercial research purposes. Physical material sent under an MTA is transferred in the form of a **bailment**, where actual legal title to the material is not transferred, but rather only a limited permission to use the material for a specified period of time.

Many of the technology transfer problems and potential dangers in an MTA are found in the section addressing rights to future inventions that may arise during the course of using the transferred materials. This section, therefore, should be negotiated with particular care. For example, if a patentable invention is developed by the party receiving the materials, who will own the rights to that invention? Should they be automatically licensed or optioned to the provider of the materials?

One MTA format used by many nonprofit organizations when transferring biological materials is called the **Uniform Biological Materials Transfer Agreement** (UBMTA). The UBMTA can be activated by a simple implementation letter between institutions that have previously agreed to abide by its terms. This saves considerable time, particularly between two nonprofit institutions that most likely will not require highly restrictive "inventions" language.[12]

Sometimes a license is granted primarily for nonfinancial reasons, with little or no remuneration paid back to the patent owner. In cases where the assignee of a patent is not the inventor, e.g., at a university, a license may be granted back to the inventor. In this case, a **Release Agreement** is signed, with or without financial terms, granting rights back to the inventor (or, in some instances, initially back to the granting agency). This might happen, for instance, at a university that is considering abandonment of an unlicensed patent, but where the inventor is willing to personally fund continued patenting and commercialization of the invention.

Consulting expertise and **know-how** rights can also be granted in technology transfer agreements, such as option and license agreements, or even be covered in their own stand-alone contracts. A **Consulting Agreement** is a contract between two or more parties whereby one party, the "consultant," agrees to provide services and know-how that are anticipated to be valuable to the other party in return for some form of remuneration (usually cash payments) for so many hours per annum.

Know-how is a form of intellectual property that is not protected by patent or copyright law, but as a **trade secret**. Some aspects of a technology are best kept secret. For example, computer companies may keep technology that becomes obsolete every few years (e.g., wiring printed circuit

---

[12] Companies that are the provider of biological materials, in this author's experience, however, rarely agree to use the UBMTA since the language concerning rights to inventions using the transferred material does not favor the provider, such as granting the provider a royalty-free, fully paid-up license to inventions that could have not been made without use of the transferred materials.

boards) as trade secrets but also patent facets that have a longer technology life, such as their computer design [2].

Although not patented per se, know-how is nevertheless often an important part of the technology transfer package being licensed. This is typically the case because a patent that refers to necessary materials and procedures may need additional know-how to fully enable the patent to become part of the optimal product. To illustrate this with an analogy, consider how a new play is often greatly improved when the author is able to attend rehearsals.

10.9 Valuation

After identifying one or more potential licensees, a common question is: what is the patent's value? The answer is, naturally, whatever the market says it is. However, unless one has previous nonexclusive licenses to use as a reference point, e.g., "cookie cutter" deals where the terms are the same each time, this answer is not very helpful.[13] There are, however, a number of approaches that may be used to address valuation and pricing of the invention.

With the benefit of hindsight, one can look at a company's financial records to determine how an allegedly infringing company may have benefited illegally from the patent rights and, thus, what a fair price would have been, particularly a fair royalty rate. However, other than in cases of patent infringement litigation where a judge is awarding damages, a licensor does not typically have the benefit of hindsight. Instead, a valuation based on assumptions, perceptions, and projections, rather than actual historical numbers, must be made.

Does cost constitute a good basis for price? With regard to pricing tangible items, this is sometimes the case. For example, the cost of gasoline is based largely on what the gas station paid. However, with patents and other intangibles, development cost is not generally a sound basis for price, and they are best priced on value, rather than on a cost basis. Products are typically priced based on total cost of the program to bring it to market, whereas patents are usually priced strictly based on their perceived value to the acquirer.[14]

One way to value a patent is to determine what it would likely cost the other party to "invent around" the patent. If the claims are narrow this might be fairly inexpensive, whereas with extremely broad claims the cost

---

[13] Patents are sometimes sold by auction, however, which can provide a market value for the patent if the number of potential buyers is large.

[14] For example, unique art when sold by the artist is priced based on the perceived value to the purchaser.

to invent around the patent might be excessively high or even undetermin-
able. This is typically a better measure of the true value of the patent.[15]

The price of a licensed patent typically consists of several different
financial terms, such as a **royalty rate** stating what portion of product
sales resulting from the invention will be paid back to the licensor. Other
terms, such as an initial cash payment (commonly called a "license closure
fee" or "up-front payment"), minimum annual payments, and payments
made upon reaching commercial milestones, are also common. Alternatively, the licensee may make one-time or periodic cash payments if the
licensor is unwilling to risk that a product may never be developed by being
compensated largely by royalties from product sales.

Most commonly, the goal of the licensor is to share in the upside
potential of the invention through a royalty rate based on net sales with
no limit, or "cap," to the amount of total royalties due over the term of the
license. The license is typically structured to profit from the license both
immediately and progressively, with an up-front fee equal to several times
the annual minimum royalties and steadily increasing milestone payments.
If substantial sales occur, however, the term with the greatest financial
impact over the term of a license agreement is generally the royalty rate.

In order to determine a reasonable royalty rate, it is important to
understand how the licensed invention will be utilized. If this is not entirely
clear, consider deferring the royalty rate negotiation until an actual licensed
product or process is introduced in the marketplace. This would require
leaving the royalty rate unspecified, with a commitment to negotiate the
royalty rate in good faith when actual income statement data become
available.[16] Most often, however, a royalty rate (or rates applying to different sales volumes) is specified in the license agreement.

A variety of ways have been used to determine royalty rates and to
otherwise estimate the value of patents. Some of these are 25% of profits
(and other "rules of thumb"), return on investment (ROI) analysis, **Discounted Cash Flow Analysis** (DCFA) (with cost of capital rates), and a
regular and established royalty. In some instances, however, it is important
to have a better understanding of the value of a patent or portfolio, such as is
the case with valuation for transfer tax purposes. Here, more than a rule of
thumb is clearly needed [14].

The rationale for the **25% rule** as a guide to determining royalty rates is
that roughly 25% of the profits resulting from the final product may be

---

[15] Furthermore, even this assessment may not properly take into account the risks involved for
the aspiring developer, e.g., that the research may not be successful, or, even if it is, that the
resulting invention may not be as effective as an in-licensed technology.

[16] In the case of government-funded institutions, a further reason for not setting a royalty rate
too early is that such an entity cannot use more than 25% of tax-exempt bond-financed space for
commercial research. Commercial research, as defined in the 1986 Tax Reform and Economic
Stabilization Act, includes "any research for which the royalty rates are set in advance" [21,
p. 13].

attributed to the licensed patent. The other 75% is considered to have come from the development and marketing of the product by the licensee. This is often used as a basis to arrive at a reasonable royalty rate. For example, if the profit margin is 20%, then a fair royalty rate might be 5%.

The use of DCFA to estimate the value of a patent involves projecting future cash flows resulting from a patented product and discounting them back to the present in accordance with the company's expected rate of financial return. The expected rate of return ("k," also known as the "hurdle rate") is company-specific and is generally higher for smaller companies, such as start-ups, since there are added risks inherent with licensing to a new business venture. A k value is ultimately determined by market pressures and thus is somewhat subjectively determined based on perceptions of risk [15].

Sometimes, computer software is used to run different DCFA scenarios many times over to get more statistically accurate numbers in what is called the **Monte Carlo Method** [16]. When assessing value with sophisticated tools such as the Monte Carlo Method, the licensee often has the advantage of more accurate assumptions and projections. The licensor, on the other hand, many times lacks the business data necessary for such an analysis to be valuable, and thus is often at a disadvantage when negotiating price.[17]

Valuation models constructed to estimate the net present value of future royalties can also be done in the format of a **decision tree**, showing the decision-making process as a pictorial representation. A decision tree consists of nodes and branches, with the numbers at the branch endpoints representing payoffs, e.g., income or loss, associated with that chain of events [17]. This analysis is useful in determining whether, e.g., the expected monetary value is higher for granting several nonexclusive licenses or for granting a single exclusive license.

Look at similar licensing deals to see what a comparable market value might be for the invention. This information can be found at a variety of service providers that obtain the information at public sources and compile it for their users.[18] Comparable deals can also be found within the license portfolio of a single institution or within the collective experience of licensing professionals.[19] Also, search the USPTO or EPO Web sites to learn how often the patent is being cited in the body of other patents since this is another indicator of the invention's relative value [18].

Several additional points should be noted concerning the valuation of patents. First, the life of the patent is an important factor to take into account during the valuation process, since a patent is a **wasting asset**. A wasting

---

[17] The price of an invention, furthermore, can also vary greatly depending on the relative strengths and styles of each negotiator.
[18] For example, for biotechnology deals, Recombinant Capital is found at *http://www.recap.com/sitehome.nsf/contact.*
[19] For example, Techno-l is found at *techno-l@lists.uventures.com.*

asset has been defined as "one that loses value, over time, which cannot be replenished, as with the loss of patent term as time passes" [14, p. 152].

Second, when assessing the probable value of the invention to a licensee, have at least a rough idea of the percentage of the total intellectual property necessary to launch the product that can be attributed to the licensed patents and know-how, i.e., not acquired from other sources. This pertains to the issue of **royalty stacking**, which is a concern of any licensee that is paying royalties to more than one licensor for the same product. For example, a contact lens manufacturer might have to pay one royalty to the owner of a patent claiming the shape of the lens, another royalty to get the rights for the hydrogel chemical composition, and yet more royalties to license patents claiming methods of polymerization and lens coating. If the total royalty burden is too great, the venture will be unprofitable for the licensee.

Finally, agreeing on the price for an invention is largely a process of negotiation, either for an exclusive license, or for the first of multiple non-exclusive licenses, which may then establish a precedent for subsequent licenses [2]. As Richard Razgaitis [15, p. 27] puts it, "[each] agreement is a snapshot in time, no two technologies are really identical, the market is never the same, and negotiators and organizations will likely be different." Thus, the terms will rarely, if ever, be exactly the same even for the same patent, with different negotiators, different companies, and/or different points in time. Valuation, nevertheless, is a valuable process and is the underlying basis with which price is negotiated. Or, as Razgaitis [18, preface] says, "valuation is an idea, pricing is an agreement."

10.10   Negotiating terms

Once an interested potential licensee has been found and at least a rough idea has been obtained of what the patent rights might be worth to that company, a common question is: how are the licensing terms negotiated? Typically, after agreeing what is (and what is not) being licensed, the first terms to be negotiated are the financial terms. The reason is that these are often the most contentious elements, so agreement here should take place first to avoid wasting time with unnecessary negotiations. The areas of greatest sensitivity and importance should be negotiated first, moving thereafter to progressively less controversial sections of the license. For example, the royalty rate will typically be negotiated before, e.g., the Books and Records section of the license agreement.

Sometimes, a prospective licensee will ask to see a "boilerplate" agreement, or template, containing the legal clauses found in most of that party's agreements. More commonly, however, one side first makes an offer of financial terms before the entire agreement is reviewed. These terms typically include some or all of the following: an up-front license fee (or license closure fee), one or more royalty rates based on a percentage of net sales, minimum annual royalties, milestone payments, patent costs, and sometimes equity ownership in the licensee as well. Particularly with a more established company, options or warrants may be proposed instead of stock. The other side then responds and the negotiation continues until either mutually satisfactory terms are reached in a **term sheet** or else the negotiation fails and no agreement is reached.

The side making the first offer can be either the licensor or the licensee and there are both advantages and disadvantages to each. The primary disadvantage of making the first offer of financial terms is that the licensee might have been willing to pay more. On the other hand, making the first offer establishes from the onset the general ballpark of what financial terms are going to be considered reasonable. Although there is a saying that "whoever makes the first offer loses," in this author's experience, if one has at least a ballpark idea of what a prospective licensee is willing to pay, making an aggressive first offer can lead to a highly successful deal.

It is important to remember that negotiators are people, and thus subject to human emotions. The increased familiarity resulting from negotiating multiple deals with the same person will make finding a set of terms that both sides are pleased with much easier. If one side is seen as being "piggish," that perception can derail a deal and, therefore, one should be sensitive concerning where this boundary lies. Moreover, during negotiations it is important to be diligently pursuing offers from other potential licensees, which can be envisioned as "keeping the line taut" [19]. This can be an effective means of applying psychological pressure to complete the agreement, without appearing adversarial.

Consider carving up the license grant into different geographic regions. For example, it might make good sense to license the Asian patent rights to an Asian company, the European rights to a European company, and the U.S. rights to a U.S. company. Furthermore, the rights to any improvements made to the invention could be a contentious issue. Typically, a licensee will want all improvements to flow into the license. Therefore, "improvements" should be defined with careful thought. For example, if a collaborating scientist participates in developing an improvement to the licensed invention, one can only promise rights to one's partial ownership of an undivided patent and, in the absence of a separate agreement with the collaborating inventor or an IIA, the license agreement should not say otherwise.

10.11   Heads of agreement

After a term sheet has been finalized by the parties, memorializing the financial consideration that will be paid by the licensee in return for the licensed patents and know-how, a **Heads of Agreement** is often negotiated. A Heads of Agreement is much more extensive than a typical term sheet and contains every article or heading that will appear in the final license agreement with a statement or paragraph stating what has been agreed upon by the parties concerning the content of each section. A Heads of Agreement typically includes [19]:

1. *What is being licensed,* which is covered under **License Grant** (e.g., an exclusive or nonexclusive license in a particular field), **Sublicense Grant** (rights, if any, the licensee has to grant sublicenses), **Territory** (geographic regions covered by the license), **Licensed Products and Processes** (specific products and processes that fall under the license), **Patents** (a list of the patents and patent applications being licensed), and **Know-How** (beyond the actual patents and patent applications, materials, trade secrets and/or other know-how that is included in the license);

2. *What is being paid in return for the license,* which is covered under **Net Sales** (how "net sales" are defined, i.e., what is excluded from gross sales to reach net sales), **Financial Consideration** (e.g., Equity, License Closure Fee, and Milestone Payments), **Royalties** (the royalty rate or rates), **Minimum Payments** (e.g., minimum royalties or annual minimums[20]), **Sublicensing Payments** (Royalties and Non-royalty Sublicense Fees), **Diligence in Commercialization** (specific timetables on the path to commercialization and other obligations the licensee has to perform beyond the financial payments due under the license), and **Patent Reimbursement and Costs** (patent costs, billing, and reimbursement);

3. *The legal framework of the agreement,*[21] which is covered under **Term** (how long the license will last), **Termination by Licensor** (e.g., not automatic, but at licensor discretion), **Termination by Licensee** (the circumstances under which the licensee can terminate the license), **Patent Infringement** (how responsibilities and obligations to pursue patent infringers and pay for patent litigation are handled), **Indemnification** (what obligations the licensee has to pay for legal defense of the licensor in the event of, e.g., a product liability lawsuit), **Dispute Resolution** (beyond resorting to litigation, other courses, such as arbitration or mediation, that will be used to resolve disputes under the agreement), **Patent Prosecution** (e.g., designation of the party primarily responsible, including rights of input by the other party), **Interest** (interest, if any, charged on overdue

---

[20] These are set at a level that will discourage "shelving," or putting the technology aside in favor of another that is not covered under your license. At the same time, they must be low enough so that, if the licensee is trying in earnest, the minimum royalty itself will not be a reason to terminate the license.

[21] This legal framework, however well conceived, may still not determine what will happen to the licensed patent rights if either the licensor or the licensee declares bankruptcy [22]. The reason for this is that the ruling court may supercede the terms of a license agreements in situations of bankruptcy and unpaid creditors.

payments), **Confidentiality** (usually a Confidentiality Agreement embedded within the license agreement), **Reporting** (obligations the licensee has to report on its progress), **Governing Law** (state or national law used to interpret the license agreement), **Books and Records** (how long they must be kept, with what detail, and licensor's right to inspect[22]), **Publications** (rights the licensor and/or licensee have to publish new findings relating to the inventions and under what circumstances), **Nonuse of Names** (whether either party can use the name of the other in, e.g., a press release announcing the deal), **Assignment** (whether the license agreement can be assigned to another party and, if so, under what circumstances), and **Compliance with Laws** (assurance that all pertinent laws and regulations will be honored).

10.12 Drafting licensing agreements

After a Heads of Agreement or term sheet has been completed, either the licensor or the licensee initiates the next phase of licensing by writing the first draft of the actual license agreement. The drafting process typically begins by using either a license agreement template as the starting point[23] or else by elaborating upon each section of the Heads of Agreement until a first draft of the entire agreement is complete.

It is typically an advantage to write the first draft of a license agreement. The rationale is that the writer of the first draft establishes the original wording and style of the agreement and can ensure that disadvantageous language is not present from the onset. The other party will likely make a counter-proposal with some liberal editing, preferably in a "red-lined" reply that shows exactly what changes were made, but the framework of the agreement is already set at this point. Changing this without completely rewriting the agreement may be difficult.

If the other party prepares the first draft, compare it very carefully to what was agreed to in the term sheet or Heads of Agreement. Even a single word change can be significant. Attorneys representing the other side may, either inadvertently or in the interests of representing their client, send a draft that does not always fully reflect the previously negotiated terms. This may reflect a perception that the entire agreement is malleable up until the point where it is actually signed.

---

[22] Large royalty-bearing licenses should probably be audited annually but smaller licenses can be audited in a periodic manner. The rationale is that for large royalty licenses, there is more money to be lost as a result of underpayment by the licensor. Therefore, incurring the cost of routine auditing makes more sense in these instances. Some licensors audit every license as a deterrent to both current and future licensees who might consider underpaying royalties [23].
[23] Examples may be found at the AUTM Web site, currently at *http://www.autm.net/index_n4.html*.

Several drafts of the license agreement are usually exchanged before the wording is finalized. This process is sometimes attempted in a single day, but much more frequently takes several weeks or months. Time, furthermore, is sometimes used as a tool to put pressure on the other side. However, keep in mind that a hastily executed document may result in ambiguous or unintended language that will need to be amended or even litigated.

Proper licensing can take a great deal of time. When licensing to a party that is not, by a judge's ruling, already infringing the patent, and thus is not under court pressure to reach a deal promptly, the time from when interest is first shown by the company to when a license is finalized can be several months. Commercial introduction of a product can take years after the license is signed, or perhaps not happen at all. In licensing to a known infringer of your patent, one must take the time to implement a licensing program and perhaps also prevail in patent litigation [14]. This can take months or even years.

10.13   Conclusion

Patent licensing is much more than just signing a contract. It is the entire rational and artistic process of valuing and conveying, sometimes very creatively, the legal right to use the invention. At times, licensing involves conveying know-how along with the patent rights or signing another technology transfer agreement, such as a CRADA, Option Agreement, or Materials Transfer Agreement to lay the groundwork for an actual license agreement. Sometimes a start-up company may be created by the licensor, initially for the sole purpose of serving as licensee to commercialize the invention.

Licensing can provide a steady stream of income with much less risk than would be the case if the patent owner were to actually develop and sell products. The licensor will typically receive a smaller share of any profits since the licensee bears most of the risk of product development and marketing. However, a single patent can be licensed to different firms if each licensee's rights do not conflict, such as granting different fields to each company.

Once the license agreement is signed, like a marriage, the formal relationship has just started. Both sides have made commitments. During the term, one may have to face decisions concerning how to address violations, or breaches, of the license agreement by the other party. From the licensor's perspective, e.g., payments need to be sent in accordance with the agreement. For this purpose, many licensors routinely audit their licensees to ensure compliance [23].

Finally, while no two deals are alike, some technology transfer arrangements are more similar than others. Lessons learned from working in the field of technology transfer can certainly be helpful when negotiating future agreements and broad categories of situations might even be defined. The following examples, while not intended to be typical, may nevertheless be instructive or otherwise helpful.

Example 1

An Associate Licensing Director at BigPharm, a large pharmaceutical company, calls the Director of Business Development at CNSBiotech, a small biotechnology firm. The larger company is interested in licensing an issued U.S. patent and a pending divisional, each relating to the diagnosis, prevention, and treatment of central nervous system (CNS) diseases. Some of the issued patent claims are blocking BigPharm's product development efforts, making it important to obtain a license. Furthermore, the divisional application would enable the pharmaceutical company to build a stronger patent wall around resulting products and methods.

BigPharm proposes to license the issued patent and divisional in the field of Alzheimer's disease. This is potentially interesting to the smaller company since its product development efforts are primarily targeted towards other CNS disorders. It is agreed, therefore, that the pharmaceutical company will draft a term sheet proposal to exclusively license the intellectual property (IP), including substantial know-how, in the field of Alzheimer's disease.

Before a term sheet is finalized, however, two additional, smaller companies, BestDrug and EZPharm, approach CNSBiotech, expressing an interest in licensing the same patent rights in the Alzheimer's field. The smaller firms are told that the licensee is already in negotiations with another company but are encouraged to make their own offers. A few days later, the three companies each make two separate proposals at the request of the licensee: an exclusive license proposal, which is preferred by each potential licensee, and a nonexclusive license proposal in the event that the patent rights become available for nonexclusive license only.

A team at CNSBiotech reviews the term sheet proposals made by each party and builds a decision tree model to evaluate the offers using DCFA. Each of the medium-sized firms offers potentially valuable cross-licensing deals as part of their proposals, but only for an exclusive license. After some consideration is given to granting three nonexclusive licenses, it is decided that granting an exclusive license to BigPharm is the logical choice since that choice yields the greatest Expected Monetary Value.

BigPharm has the most advanced development program in the Alzheimer's field, owns or has licensed several synergistic patents, and is offering the most guaranteed cash payments. Some of the payments are offered to be made in the first few years of the license whereas others will be payable upon reaching specified development milestones. Another influencing factor is that the large pharmaceutical company is further along than the other two firms both in its internal product development program as well as in having assembled the other necessary patent rights. BigPharm is also offering a very large guaranteed up-front licensing fee as part of the consideration for an exclusive license.

As terms continue to be discussed with the three potential licensees, CNSBiotech receives an unexpected letter from the USPTO, proposing a possible interference between the issued U.S. patent and another pending U.S. patent application filed by an unnamed third party entity. The biotechnology company promptly discloses to BigPharm, BestDrug, and EZPharm that an interference proceeding may potentially be declared by the PTO. This news, however, does not lessen their interest in licensing the patent.

The deal that is finally reached includes an option to license the other, non-CNS, applications of the technology in return for additional payments. The license terms also include the license closure fee (paid in installments) and even larger potential milestone payments. The royalty scheme guarantees a certain royalty rate regardless of whether the licensed patents are utilized in a final product due to the know-how that will be provided even if, e.g., the patent interference is lost.

Example 2

A Professor of Physics, "Dr. Nanotech," telephones her university's technology transfer office expressing interest in forming a start-up company to commercialize her nanotechnology patent application filed and owned by the university. The pending claims of this patent are very broad, with foreseeable applications for several different products and markets. Dr. Nanotech's coinventors (also employed at the university) are interested in forming a start-up company as well, so a meeting is arranged between the inventors and the director of the technology transfer office. Given that the invention is a "platform technology" and the inventors agree to hire professional management, it is agreed that a new company will be formed and a license agreement negotiated with the start-up.

The inventors work with a local venture management group to provide management and some initial seed capital for the new company, "SmallCo". The CEO of the venture management team is selected to be the first President of the start-up company, and this individual, in collaboration with a local law firm, responds to a term sheet proposal made by the university.

In addition to a modest royalty rate, the license agreement with SmallCo calls for payments to be made to the university upon the company reaching several research, development, and product launch milestones. Furthermore, under the agreement, the university receives a minority stock-ownership in the firm, with such founders' equity to be nondilutable until a defined amount of capital is invested in the company.

A milestone payment in the license agreement calls for a fee to be made to the university upon the issuance of the licensed patent in the U.S. During patent prosecution, however, the originally filed patent application licensed to SmallCo is split into several separate divisional applications. When the first divisional is issued in the U.S. with only some of the original claims, however, a disagreement arises over whether that particular milestone has been reached.

A compromise is agreed to where the payments will be made in install-ment parts as each divisional application issues. This accomplishes three goals: first, costly litigation is avoided by both parties; second, the start-up is able to preserve its limited funds; and third, the university receives modest payments and has the opportunity to eventually be paid the entire mile-stone fee.

Example 3

A retired corporate engineer, "Mr. Fuzzy Logic" develops an image pro-cessing software invention on his home computer system. Upon the advice of a former colleague, the engineer discusses his invention with a reputable technology broker for evaluation of the invention as a potential licensing opportunity. Although the broker sees the invention as having too limited a market for them to devote the resources to work on it, some helpful feed-back is provided to Mr. Logic.

Using the Internet and his previous experience in industry, Mr. Logic makes a list of several firms that might be potential licensees. After contact-ing the President of one of the firms, "DigiQuip", both through the mail and over the phone, the company expresses interest in potentially licensing the software for bundled sale with its equipment. DigiQuip and Mr. Logic sign both a Confidentiality Agreement and a Materials Transfer Agreement provided by an attorney hired by the inventor. DigiQuip evaluates the software on its own equipment and the parties subsequently decide to execute an Option Agreement until significant further market research is done.

An Option Agreement is negotiated whereby DigiQuip will be granted an option to obtain an exclusive, worldwide, license to the technology for terms to be negotiated, but with a royalty rate not to exceed 10% of the percentage difference between the price of the bundled product and the equipment by itself. Furthermore, the license grant will cover not only the image processing software and technology but also improvements made by the inventor during the life of the license. The company will also have the option of filing patent applications in collaboration with the inventor but at the company's expense.

The Option prohibits the inventor from entering into consulting agree-ments with new competing companies but exempts any previously existing relationships. Know-how is included in the arrangement and Mr. Logic must commit to include any improvements made during the option period in the eventual license.

In consideration for a 12-month Option, a modest option fee is agreed upon that is creditable against the license fee if the option is exercised. The term of the agreement, however, is a major point of contention since Digi-Quip wishes to test-market the software for several months. Eventually, an option term of 12 months is agreed upon with the possibility of extending this term by payment of an extension fee.

References

1. Dworetsky, S.H., in the satellite program "Intellectual Property Issues In Structuring Deals and Drafting Agreements 2002: Strategic Patenting," June 2002.
2. Jolly, V.K., *Commercializing New Technologies: Getting from Mind to Market*, Boston, MA, Harvard Business School Press, 1997.
3. Meg Wilson at *http://www.pax.co.uk/ttdefine.htm*.
4. LES: The Basics of Licensing, 1988.
5. *http://www.geocities.com/MotorCity/Lot/3248/hist01.htm*.
6. "Pre-Production Investment and Jobs Induced by MIT Exclusive Patent Licenses: A Preliminary Model to Measure the Economic Impact of University Licensing" at *http://www.autm.net/pubs/journal/95/PPI95.html*.
7. Yoichiro Iwasaki, Licensing: A Tool to Expand Business, *les Nouvelles*, Vol. XXXVI, No. 1, March 2001, p. 5.
8. von Belvard, P.R. and Buechel, K.F., Inventors Need Patent, Licensing Help, *les Nouvelles*, Vol. XXXIII, No. 3, September 1998.
9. *http://www.autm.net/index_ie.html*.
10. *http://www.t2s.org/abou.html*.
11. Thayer, A., Closing the Venture Gap, *Chemical & Engineering News*, Vol. 78, No. 33, 2000.
12. De Corte, F., Interaction between Industry and Universities, *les Nouvelles*, Vol. XXXVI, No. 1, March 2001, p. 13.
13. Personal communication with G. Michael Alder.
14. Bramson, R.S., Rules of Thumb: Valuing Patents and Technology, *les Nouvelles*, Vol. XXXIV, No. 4, December 1999, p. 149.
15. Razgaitis, R., Pricing the Intellectual Property Rights to Early-Stage Technologies: A Primer of Basic Tools and Considerations, *AUTM Manual*, Part VII, Ch. 4, Association of University Technology Managers, 1994.
16. Razgaitis, R., *Early-Stage Technologies: Valuation and Pricing*, Wiley, New York, 1999.
17. Anderson, D.R., Sweeney, D.J., and Williams, T.A. *Quantitative Methods for Business*, 4th ed., 1989.
18. Rivette, K.G., and Kline, D., *Rembrandts in the Attic: Unlocking the Hidden Value of Patents*, Boston, MA, Harvard Business School Press, 2000.
19. Personal communication with Edward R. Gates.
20. *http://www.lesi.org/level2a/conduct.htm*.

21. Stevens, A.T., Finding Comparable Licensing Terms, *AUTM Manual*, Part VII, Ch. 5, Association of University Technology Managers, 1994

22. DeMeo, K.L., License Agreement in Bankruptcy, *les Nouvelles*, Vol. XXXIV, No. 1, March 1999.

23. Goldscheider, R., in the course "International Licensing and Negotiation for the Technology Manager," December 1991.

Index

Printed in the United States
by Baker & Taylor Publisher Services

Printed in the United States
by Baker & Taylor Publisher Services